D1712936

The Art of Managing Longleaf

The Art of Managing Longleaf

A Personal History of the Stoddard-Neel Approach

LEON NEEL
with Paul S. Sutter and Albert G. Way

The University of Georgia Press
ATHENS & LONDON

Publication of this book was made possible in part by
a subvention from the Joseph W. Jones Ecological Research Station
at Ichauway

All photographs, unless otherwise noted,
are courtesy of Leon and Julie Neel.

Map on page 40 by David Wasserboehr.

Athens, Georgia 30602
www.ugapress.org

Designed by Erin Kirk New
Set in 10.7/14 Minion
Printed and bound by Thomson-Shore and Pinnacle Press

The paper in this book meets the guidelines for
permanence and durability of the Committee on
Production Guidelines for Book Longevity of the
Council on Library Resources.

Printed in the United States of America
13 12 11 10 09 C 5 4 3 2 1

Library of Congress Cataloging-in-Publication Data

Neel, Leon, 1927–
The art of managing longleaf : a personal history of the Stoddard-Neel Approach / Leon Neel with Paul S. Sutter and Albert G. Way.
p. cm.
Includes bibliographical references and index.
ISBN-13: 978-0-8203-3047-1 (hardcover : alk. paper)
ISBN-10: 0-8203-3047-7 (hardcover : alk. paper)
1. Longleaf pine—Southern States—History. 2. Forest management—Southern States—History. I. Sutter, Paul. II. Way, Albert G. III. Title.
SD397.P59N44 2010
634.9'751—dc22 2009029173

British Library Cataloging-in-Publication Data available

To my wife,

Julia Greene Neel,

and our daughters,

Julie and Susan

CONTENTS

ILLUSTRATIONS

The Art of Managing Longleaf

INTRODUCTION

Forestry beyond One Generation

by Paul S. Sutter and Albert G. Way

We first met Leon Neel in, of all places, a parking lot. It was a radiant morning in downtown Thomasville, Georgia, in the spring of 2004. Just two miles west of us lay some of the most beautiful land in the southern coastal plain, land that we knew contained prime examples of an endangered longleaf-grassland biome that once stretched for tens of millions of acres in all directions. We had been invited by the Joseph W. Jones Ecological Research Center at Ichauway to discuss undertaking an oral history project centered on Leon's life and work with the southern longleaf pine. Our liaison, Kevin McIntyre, had carefully described what they had in mind as we drove from the Jones Center, located in Baker County south of Albany, to Thomasville, about an hour to the southeast. Kevin also made clear that several other candidates were vying for the project and that gaining Leon's trust would be critical to our chances. So we were eager to meet Leon and make a good impression. We were also keen on seeing those magnificent longleaf woodlands.

As we chatted with Kevin in the parking lot, Leon drove up in an old long-bed Chevy, its cab littered with the debris of doing business in the woods. He sprang from the truck, greeted each of us with a firm handshake, and after a bit of small talk, suggested we head out of town. We were clearly not the first people to be interested in his work, but Leon is a gracious man, and he will talk eagerly to almost anyone about the landscape to which he has devoted his life. And as we listened, we quickly realized that this project, if we were lucky enough to get the assignment, would be a special opportunity.

You do not have to spend much time with Leon Neel in the longleaf woodlands that he has managed to learn that they are filled not only with ecological diversity and a potentially sustainable supply of timber but also with stories. While Kevin had made it plain that the Jones Center hoped such an oral history project would capture a detailed rendering of Leon's forestry practices in his own words, he also intimated that the stories Leon had to tell about the woods and their inhabitants were as rich as his applied ecological insights. After several years of working with Leon on this project, we are now convinced that those stories are in fact part and parcel of his ecological management approach. Leon has taught us many things, but one of the most important is that if you do not have stories to tell about the lands you work and love, how can you hope to protect them? This book, then, is about the history and tenets of Leon Neel's approach to land management and the stories that inform it.

Our first stop that morning was Greenwood Plantation. It was, in a word, breathtaking. Leon refers to Greenwood often in this book, and from a longleaf conservation standpoint it is arguably the best known of the Thomasville-Tallahassee properties. Like so many former antebellum plantations in the Red Hills, as the surrounding region is known, Greenwood passed from southern patrimony to northern money after the Civil War and eventually landed in the hands of Oliver Hazard Payne, a Cleveland businessman with Standard Oil connections. At his death in 1917 Payne left Greenwood to his nephew Payne Whitney, who in turn left it to his son, John Hay "Jock" Whitney, in 1927. It has been in the Whitney family ever since. The "Home Place" of Greenwood, as the plantation's central property is called, is about four thousand acres, a quarter of which is aptly referred to as the "Big Woods."

There are many patches of forest called the Big Woods in the South, and they are all special, even those mythical big woods of William Faulkner's fiction. The Big Woods of Greenwood, however, bear little resemblance to the once-prodigious bottomland forests of Faulkner's Mississippi Delta. For one thing, these Big Woods still exist. For another, they are upland pine savanna rather than river and swamp hardwood. Frequently called "the best of the last," Greenwood's thousand-acre Big Woods is arguably the most remarkable stand of longleaf pine grassland remaining in the southern coastal plain. It is an old-growth, multiaged woodland with some longleaf trees more than four hundred years old, their flattened crowns betraying their age. The understory of the Big Woods is burned frequently—at least every two years—and once grown back, its diverse mix of native warm-season grasses, legumes, forbs, and shrubs sparkles when the sun hits it right. The Big Woods is a full-color specter of the vast and legendary coastal plain forests now preserved mostly in old sepia-toned photographs. For its existence we can partly thank Leon Neel and his mentor, Herbert Stoddard. Leon has helped to manage Greenwood for over fifty years, and Stoddard worked it for twenty-five years before that. It was their primary laboratory and show place, and Leon is justifiably proud of it.

Herbert Stoddard developed, and Leon Neel perfected, an ecological approach to forestry practice before anyone had coined the phrase "ecological forestry." Much like present-day ecological foresters, Stoddard and Neel sought to mimic natural-disturbance events and other ecological processes through controlled burning and sustainable timber harvests, while simultaneously maintaining and enhancing the longleaf woodlands' biological function and diversity. The Stoddard-Neel Approach, as their method came to be called, is more a set of principles than a textbook forestry method. As a result, those who like their forestry methods rendered in neat, abstract formulas have often expressed frustration with the approach. Nonetheless, when carefully practiced over the long term, Stoddard-Neel is one of the few forest management techniques that ensures the survival of ecological integrity and function from generation to generation while still allowing for substantial timber harvests. Continuity, then, is at the core of the Stoddard-Neel Approach—continuity of diverse biological life, continuity of timber resources, and continuity of human pleasure in the aesthetic beauty of places such as Greenwood.

Although neither of us knew that much about longleaf woodlands back in 2004, we had done some preparatory reading for our first meeting with Leon. We knew a little about how the longleaf-grassland system worked, about its history, and about the various scientific arguments for why it needed to be preserved. But neither of us was quite prepared to make sense of the unusual beauty of Greenwood. Yet that aesthetic reaction is the foundation of the Stoddard-Neel Approach. Many foresters are quick to dismiss aesthetics as a proper measure of good forestry, or they are uncomfortable with a set of values that seems not only far removed from the efficient production of timber but sometimes even hostile to it. Leon Neel, and Herbert Stoddard before him, however, used the look of the woods as a gauge to measure their health. Indeed, we were not long into our initial tour of Greenwood when Leon educated us on the way a longleaf woodland should look. First, he instructed, a healthy longleaf woodland allows one to see a great distance through the trees but also always to see trees. As he makes clear in the memoir that follows, that long look through the forest, which early quail hunters in the region prized, is an important metric of several critical functional aspects of longleaf ecology and management. Second, Leon was quick to point to the many small patches, or "domes," of regeneration that dotted the understory. Those were the future of the forest, he insisted, as important as the gnarled flattops to his practice of forestry beyond one generation. When inspecting the woodlands he manages, Leon is quick to admire good, thick patches of regeneration in the small openings made by his careful forestry practice, and we have come to take joy in them too. Third was the diversity of the understory as it existed across a landscape gradient defined by altitude and moisture. We stopped frequently to admire the seasonal blooms of orchids and other wildflowers and to note how the dry uplands gave way to thicker growth in the hardwood drains that ran through the Big Woods. While Leon insists that you can gauge the health of a longleaf woodland by how it looks, he also has taught us that no two healthy longleaf woodlands look exactly alike. Indeed, part of the aesthetic joy to be taken from these landscapes comes precisely in recognizing how geology, soils, microclimates, moisture gradients, and disturbance histories sculpted them into a once-vast mosaic. Many of Stoddard's and Neel's most important ecological and management insights came from plumbing their sense of the diverse beauty of these woodlands. This is one reason why Leon has consistently insisted that the Stoddard-Neel Approach is an art, not an exact science.

The Stoddard-Neel Approach took shape in what was once one of North America's signature biomes: the longleaf pine savannas and woodlands of the southeastern coastal plain. Its original range extended from southern Virginia through the Deep South to east Texas, forming a ninety-million-acre bow of land staggering in its biological diversity. Like the prairies of the American Midwest, it is a historic plant association that now exists only in small patches and fragments. Within this vast bioregion, the specific area surrounding Thomasville, Georgia, and Tallahassee, Florida, loomed large for Stoddard and Neel. Known as the Red

Hills, this section mostly escaped the wave of industrial timbering that swept through the coastal plain at the turn of the twentieth century. As a result, when Herbert Stoddard first came to the Red Hills in 1924, he encountered large contiguous blocks of functioning longleaf woodlands that few other parts of the coastal plain could match. Within these woodlands, he developed an innovative model of conservation management.

Many of the stories Leon told us, both as we toured Greenwood that first day of our acquaintance and over the subsequent years of working with him, were about Herbert Stoddard and his pioneering work. Indeed, to know Leon Neel is to know about "Mr. Stoddard," as Leon still respectfully refers to his former boss and mentor. Herbert Stoddard can justifiably be called the Aldo Leopold of the Southeast. While he never produced a volume as elegant as Leopold's *A Sand County Almanac*, he reached many of the same conclusions about the need to think ecologically and act ethically in relation to the land. Indeed, Stoddard and Leopold were good friends and mutual influences, and Leopold was quick to give credit to Stoddard for land management innovations often ascribed to Leopold. But compared with trained conservation experts such as Leopold, Stoddard's background was striking in what it lacked. Raised in a family of modest means, he had no formal education beyond primary school. Even so, by the end of his career, he was widely recognized as the "father" of wildlife management and as a pioneering if controversial figure in the nascent field of fire ecology. To those legacies we hope to add "ecological forester."

Herbert Stoddard was born in 1889, the second child of a working-class family in Rockford, Illinois. In 1893 Stoddard's stepfather gambled on the central Florida land boom, moving the family south to have a go at growing oranges. Stoddard's memories of his early childhood years spent wandering and working in Florida's inland longleaf forests stayed with him for a lifetime. When the family's immature orange grove succumbed to a big freeze in 1895, young Herbert started spending time with local cattlemen and gained a nascent understanding of how fire worked in the longleaf system. He would continuously draw on that early experience in the Florida backwoods in his later wildlife work.

After seven hard years on the Florida frontier, his family moved back to Rockford in 1900, where Stoddard soon tired of school and took off to work on his uncle's farm in Prairie du Sac, Wisconsin. There he met a local taxidermist named Ed Ochsner, and he began an apprenticeship that eventually led to taxidermy jobs at the Milwaukee Public Museum and the Field Museum in Chicago. Stoddard's specialty was birds, and he soon made the necessary connections through ornithological fieldwork to land an unusual position with the U.S. Bureau of the Biological Survey (BBS) in 1924. Landowners in the Red Hills had been negotiating with the BBS about launching a field investigation of the life history of the bobwhite quail in their area. They promised to fund the study if the BBS would handle the administrative work and appoint a suitable investigator. After a nationwide search, the BBS offered the job to Stoddard. The venture became known as the Cooperative Quail Investigation (CQI), a name that reflected the public-private nature of the initiative. Stoddard was charged with establishing a research station in the Red Hills to study the population dynamics of bobwhite quail and to implement strategies to ensure their increase, largely for the purposes of recreational hunting. The results of his pioneering study, however, would extend far beyond the fate of that one bird and its usefulness to wealthy hunters.[1]

When Stoddard arrived in the Red Hills, he entered a social landscape of astonishing wealth and undeniable poverty. As in the rest of the South, tenantry and sharecropping were staples of the agricultural economy in the Red Hills, and the crop-lien system was a firmly entrenched economic reality, keeping landless farmers in a cycle of debt. Landowners rented out land, implements, and fertilizer in exchange for a portion of a crop that rarely made, and the debt piled high for most tenants. Economic and social disparities between the races were defining features of the Red Hills, as they were throughout the South. White sharecroppers were present, as were black landowners, but the converse was far more common. This was the social landscape of the Red Hills in 1924, though there was one important distinction between it and much of the rest of the South.[2]

Over the previous several decades, the Red Hills had become the winter home of some of the nation's wealthiest industrialists and the birthplace of the southern quail preserve. In the closing days of Reconstruction, Thomasville's leadership began positioning the town to capture part of a growing market of health-seeking tourists from the North. Soon, the Red Hills became known for its well-drained rolling hills and its large expanses of open longleaf pine woodlands. This was just the sort of place recommended by physicians of the day to combat tuberculosis and many other so-called urban diseases. The resinous needles of pine trees, in particular, were thought to filter air in ways that made it particularly salubrious, and vigorous activity among these trees was a familiar prescription for remedying poor health. In the South, the longleaf pine was supreme among the pine species, and few places could match the quality of the Red Hills' longleaf woodlands. During the 1880s and 1890s, then, northerners in growing numbers wintered in Thomasville, occupying grand new hotels like the Piney Woods and Mitchell House, and they increasingly took to the fields and forests in search of healthy recreation.

Sport hunting bobwhite quail became especially popular among this wintering set, as quail make for an exciting hunt. Hunted over the point of well-trained dogs (usually pointers or setters), a quail covey usually lies still until flushed, when a dramatic eruption of birds tests the accuracy of even the finest shooters. Much of the rural South harbored ideal environmental conditions for bobwhite quail in the years following the Civil War. Sharecropping and tenantry may have been economically degrading for the human inhabitants of the South, but the resulting land-use patterns provided an ecological bounty for such wildlife as quail. The bobwhite quail is a ground-nesting bird that thrives in the patchy early successional habitats found in small-scale, preindustrial agricultural landscapes. The field edges and hedgerows of the tenant landscape provided perfect nesting sites and escape cover from predators, and the laid-by and abandoned fields produced a similar mix of food-producing vegetation. In the Red Hills the remaining long-leaf forest and its open understory of early successional plant life, the result of regular local burning, extended the quail range into the woodlands. It also produced an open pine woodland aesthetic that many prized as an environment in which to hunt. While locals hunted quail with little more than a dog and gun, northern visitors developed an elaborate hunt with specially designed off-road carriages, dog pens, gun boxes, and a platoon of dog handlers and servants.[3] They grew so enamored with the sport and the place that many returning visitors began to purchase land outright, converting former plantations into a series of exclusive winter retreats. Between 1880 and 1920 northerners bought nearly three hundred thousand acres of land in the Red Hills, and they established the area as the center of a new trend in regional land use—the southern hunting preserve. But after several decades of heavy shooting and changing land-use patterns, winter residents noticed quail numbers declining by the early 1920s, and, with little idea of the cause, they turned to the Biological Survey for answers. The BBS fatefully hired Stoddard—the young taxidermist, amateur ornithologist, and museum field scientist—to study the situation and provide these wealthy landowners with some answers.

The field of wildlife management was all but nonexistent when Herbert Stoddard came to the Red Hills. Beyond the anecdotal knowledge of hunters, some of it accurate and some fanciful, little was known about the life history of bobwhite quail and other game species. Stoddard's study would change that. In the four-year lifespan of the CQI, he developed many of the field techniques that wildlife scientists would use for decades, and he placed the anecdotal knowledge of hunters under the scrutiny of the scientific method. He monitored and described typical quail behavior and social structure throughout the seasons, determined their favored foods, experimented with seed collecting and methods for planting their favorite food plants, and distinguished between their perceived and actual predators. One of Stoddard's major conclusions was that the fate of quail and other wildlife rested with the quality of their habitat rather than strict bag limits. He linked quail with the landscape components of small-scale agriculture, and he encouraged landowners to maintain the field edges, hedgerows, and weedy fields that made the Red Hills such a haven for quail in the first place. The book that resulted from his study, *The Bobwhite Quail: Its Habits, Preservation, and Increase* (1931), became a foundational text for the field of wildlife management, and it is still often referred to as "The Quail Bible." Much of Stoddard's study focused on quail biology in agricultural landscapes, but actively farmed lands covered only a small portion of the Red Hills. He also ventured into the longleaf woodlands, and it was there that he pieced together the early tenets of the Stoddard-Neel Approach to forestry. In the process, he helped to shake up the forestry establishment.[4]

If the world of science knew little about wildlife population dynamics in the 1920s, it knew even less about the mechanics of the longleaf-grassland system. Some suspected fire was an important influence in its evolutionary past, but no scientist had yet ventured much further than to guess what that influence was. Stoddard's work on the bobwhite quail, though, helped to usher in a new understanding of the ecology of the upland coastal plain. He, along with a few botanists, foresters, and land managers scattered across the region, came to see longleaf woodlands as dependent on frequent fire for their perpetuation. This realization deflected conservation science in the region onto a new trajectory, one that led to our current appreciation of longleaf as a "fire forest."[5]

The ecology of the longleaf pine–grassland system is a product not only of the elements—earth, air, water, and fire—acting across deep time, but it is also the product of human interventions across millennia. The once-vast longleaf woodlands of the Holocene era coastal plain have never been a wilderness, if we mean by wilderness a landscape unmarked by the workings of human culture, and the Stoddard-Neel Approach does not treat them as such. Rather, SNA is a new iteration on a long tradition of human management with nature.

First, there was water. The southern coastal plain has been submerged—partially or wholly—many times since the Cretaceous period, and the long geological fall line that today divides it from the hilly Piedmont once marked the shores of the Atlantic Ocean and the Gulf of Mexico. During the last glacial period about eighteen thousand years ago, the shoreline extended hundreds of miles beyond our current shores, and only receded to its present position when a warming trend released glacial water to flow back into the seas. Water, then, gave the coastal plain its flat topographical character. Once the coastal plain assumed the boundaries we recognize today, water continued to shape the structure of its plant communities, often in tandem with fire.

Then there was earth. The soils that emerged on the coastal plain varied as a result of this long depositional history, though most areas were made up of coarse, acidic, well-drained sandy soils.[6] In the Red Hills, however, as the name would suggest, the soil formed as fertile red clays of nonmarine deposits and rolled from hill to dale as a result of fluctuating sea levels. These red clays held water and grew lush, diverse plant life, and they would distinguish the longleaf-woodlands of the Red Hills from those of other regions with differing soil profiles.[7]

Wind, too, has had an important influence on the coastal plain and its longleaf woodlands. When we first visited Greenwood with Leon, we enjoyed a gentle breeze that made the pines sway and whisper, one of the signature aesthetic experiences of being in a healthy longleaf woodland. But this landscape has been historically shaped by violent winds as well, in the form of hurricanes and sometimes tornadoes. Winds of such force can create large-scale disturbances with long-term ecological legacies, and, as we will see, the management challenges of such disturbances proved to be powerful catalysts in the emergence of the Stoddard-Neel Approach.

These natural elements—waters, soils, winds—gave shape to the southern coastal plain, but fire, more than any other force, determined the composition of the upland longleaf pine–grassland system. At times a catastrophic force, but more often a low-level and replenishing one, fire has been the single most important factor in the evolutionary history of the region's woodlands. Fire's long history on the coastal plain begins with lightning. The South experiences an average of sixty to eighty thunderstorms per year, most of which are concentrated in the summer months. When coupled with the right fuel on the ground, a single thunderstorm could start multiple fires across the coastal plain landscape. Prior to the landscape fragmentation caused by extensive deforestation and agricultural settlement, those fires often moved across tens of thousands, or sometimes even millions, of acres. The southern coastal plain developed into what ecologists sometimes call a fire climax, a landscape that not only supported natural fires but also increasingly needed them.[8]

Considerable debate exists in the scientific community about whether or not lightning alone can explain the longleaf-grassland system's dependence on fire. It was undoubtedly a formative force, but humans too have been carriers and spreaders of fire since arriving in the coastal plain over ten thousand years ago. Native Americans used fire for a variety of reasons, most often to attract game animals, to control pests, and to make the forests more manageable for everyday life. Early European and African settlers also recognized the advantages of fire in the longleaf forests, and their rural descendents continued burning the woods for generations. Indeed, some authorities argue that the longleaf-grassland system as we know it is little more than five thousand years old, which would suggest that human-set fire had as much influence on its development as lightning. In contrast, others point to fossil pollen evidence that shows the existence of pine and oak savanna long before humans appeared on the coastal plain, or to a prehuman faunal record that contains a number of creatures with morphological adaptations indicative of a savanna landscape in the region during previous interglacial periods. There has been, in other words, a long and vigorous debate about the age of this particular ecological system and its most active evolutionary agents. We do not seek to resolve this debate here. What we can say for sure, though, is that both natural and anthropogenic fire eventually converged as pine-grassland environments came to dominate the coastal plain during the Holocene period, and that the longleaf pine–grassland system adapted to frequent fire as a result. This historical frequency of fire, in turn, is now only duplicated in active management systems such as the Stoddard-Neel Approach.[9]

This mix of natural and human-set fires shaped a forest in which virtually all native plant and animal species have some adaptation to fire, and many are entirely dependent on it for their survival. Longleaf pines are the dominant tree species on the upland coastal plain. While they do not have serotinous cones (cones that require fire in order to open up and release seed) like some other fire-adapted pine species, they do require bare mineral soil of the sort exposed by fire in order for their seeds to germinate. Moreover, longleaf pines have adaptive strategies for surviving fires during the early stages of their development. After germination, while the longleaf establishes its long tap root below ground, up above it spends anywhere from three to fifteen years in a grass stage with long needles that protect the terminal bud from fire. During this grass stage, longleaf are often, for the novice at least, indistinguishable from the bunch grasses that populate the understory. This ability to survive fire in the seedling stage gave longleaf pines a distinct advantage over other tree species in such a fire-prone region. Other southern pines such as loblolly, slash, and shortleaf can tolerate a cool fire after several years of growth, but not while they are seedlings. Nor can most hardwood species survive frequent fire. Frequent fires, then, kept the spread of longleaf's competitors in check. But without frequent fire, the understory of a longleaf forest grows up into a thick rough, and, depending on local conditions, successional hardwoods such as blackjack oak, bluejack oak, red oak, and sweetgum encroach on the uplands, eventually crowding out the longleaf and understory grasses and setting the landscape on a different developmental trajectory. The longleaf-grassland system that dominated much of the coastal plain for thousands of years, then, was an artifact of fire.[10]

One of the defining grasses of the longleaf system, wiregrass, is also heavily dependent on fire for its existence. A highly combustible species, wiregrass—in combination with resinous longleaf needles—plays an important part in the longleaf woodlands of the Red Hills region. It is a dry, fibrous grass, and even when it greens up during the summer, it actually has as many dead as living blades, making it an ideal carrier of fire. Today, wiregrass is increasingly rare, a marker of unplowed and otherwise undisturbed land. When agriculture spread across the coastal plain, much of the region's wiregrass was plowed up, and because wiregrass requires a growing-season fire to germinate, it can be very difficult to reestablish after tillage. As a carrier of fire, wiregrass is highly desirable from a management standpoint, though it is not essential for the maintenance or restoration of longleaf forests; indeed, there were parts of the historical longleaf belt in which wiregrass was not a common understory component, and today a considerable amount of longleaf restoration proceeds on lands without a functional wiregrass understory.

Wiregrass and other understory plant species were so important to the integrity of the original longleaf range that many observers have described the system as more of a savanna ecosystem or forested grassland. Indeed, if we use current vegetative classifications as our guide, most longleaf landscapes would not have sufficient tree density to be considered forests. They are more accurately called woodlands, a term that we have tried to use consistently in this introduction. While we use terms such as forest, woodland, savanna, and grassland interchangeably throughout the rest of this book, we do recognize that there are important technical distinctions between them. The important point regarding the historical aesthetic of the longleaf system is its relative openness. With frequent fire arresting most of the midlevel vegetation, this openness permitted sunlight to stream through the canopy to encourage vigorous understory growth. Gaps in the canopy, caused by natural tree mortality or disturbance events, allowed regenerating longleaf saplings to replace older trees, resulting in a multiaged forest at various stages of growth and maturity. The density of the canopy was important to the maintenance of the system. An overly dense canopy deprived the grassland of necessary sunlight. An area sparsely populated with longleaf, on the other hand, lacked the pine-needle fuel necessary to carry ground fires and thus maintain the system. Despite the observations of some early coastal plain travelers who found forests so dense that they had to watch their shins while on horseback, these places were likely rare. The spatial density of these forests more often provided a large field of view, an aesthetic that Stoddard, Neel, and many of the region's residents and visitors came to value as distinctive—even if many early visitors found the forests to be barren.

The longleaf itself reigns supreme among tree species in this system and is essential to its integrity, but the system's floral and faunal diversity lies in the understory. In fact, longleaf pine woodlands may well be the most diverse North American ecosystem north of the tropics. Its small-scale diversity—the number of different understory species found in particular small-scale plots—rivals the most diverse systems on earth, and there is a high degree of endemism—plants found only in the region—as well.[11] On remnant longleaf lands that retain their historic ecological integrity, the understory typically contains 150 to 300 species of ground cover plants per acre, about 60 percent of the amphibian and reptile species found in the Southeast—many of which are endemic to the longleaf forest—and at least 122 endangered or threatened plant species. Habitat loss and fragmentation have placed keystone species like the red-cockaded woodpecker on the federal endangered species list and in places made the gopher tortoise threatened. Other important markers of forest health that are in peril are the flatwoods salamander, the eastern indigo snake, the striped newt, and the Florida pine snake.[12] All of this diversity hangs on in places like Greenwood Plantation's Big Woods. And had Herbert Stoddard and Leon Neel not actively burned these lands as part of their management, much of that diversity would likely be gone.

Into this remarkable region came Herbert Stoddard in 1924, and he remained actively engaged with this landscape until his death in 1970. In that time span, he became a longleaf polymath. He was expert at the so-called applied sciences of wildlife management, agriculture, and forestry, and despite not having any formal education he never shied away from the theoretical questions of ecology. We likely would call him a conservation biologist today, but Stoddard himself preferred the title of land manager. On the Red Hills quail preserves, he could shape a piece of land to his liking, usually with a nod to the bobwhite quail, but almost always with a larger concern for the overall biological integrity of the system. After the initial quail investigation ended in 1929, some preserve owners lured him away from the Biological Survey with the promise of wildlife consulting work in the Red Hills and other hunting preserve regions in the South. In addition, one generous landowner, Lewis Thompson, gave Stoddard Sherwood plantation, the one thousand acres of land he had used as a home base during the quail investigation. Stoddard then established the Cooperative Quail Study Association (CQSA) in 1931, a private group of dues-paying landowners who funded Stoddard's continued research and paid him as a consultant for their quail preserves. Stoddard met Ed and Roy Komarek in 1933, and he hired Ed as an assistant the same year, which would prove to be an important step in sustaining the Red Hills as a center for ecological research.[13] Not far away, Leon Neel was growing up on the land in the midst of a deep economic depression.

Stoddard's turn as a forester began in 1941, after a powerful hurricane felled large volumes of timber in the Red Hills and the U.S. government began to request larger timber harvests for war materiel. Stoddard had advised preserve owners to thin their woodlands for years, and he seized on the hurricane and the war as opportunities to improve the region's woodlands both economically and ecologically. The preserve owners were reluctant to cut any timber at all. The turn-of-the-century southern timber boom had left a devastated coastal plain landscape, and the images of desolate clear-cut lands still loomed large in their minds. But with the reassurance that Stoddard would direct the operations, some owners agreed. He brought in salvage logging crews to extract the wind-damaged timber, and he also marked some individual standing trees for harvest. The effect was to open up dense stands of timber, thus encouraging the vigorous growth of ground cover that so many wildlife species depended on. Once word spread about the efficacy of Stoddard's forestry operations, the close management of forest resources became commonplace on most southern hunting preserves.

It was at this point, just after World War II, that Leon Neel entered the picture, having graduated from the University of Georgia's forestry school and returned to his hometown of Thomasville. As Stoddard moved into the timber consulting business more aggressively, Neel assumed the role of trusted assistant, and, over the subsequent years, he became more like an equal partner, particularly as Stoddard aged and proved less capable of working in the woods. Much of the memoir that follows charts their relationship during these early years and offers a compelling story of trust and friendship in their time spent together in the woods.

As Stoddard and Neel perfected their approach to forestry and land management, they also teamed up with others in the area to form an impressive cohort of independent scientists and naturalists, a group that included Henry Beadel, Ed Komarek, and Roy Komarek. With Herbert Stoddard at its center, this group spoke throughout the 1950s of establishing an institutional home for their accumulated knowledge about the longleaf woodlands. They did so in 1958, when Beadel deeded over his Tall Timbers plantation as a land base for what became Tall Timbers Research Station, a privately funded institution with a very public mission. Under the leadership of Stoddard, Beadel, the Komareks, and Leon Neel, Tall Timbers sponsored pioneering conferences on fire ecology and ecological land management, published the proceedings of these meetings as well as many other papers and pamphlets, and conducted research on a wide range of questions about coastal plain ecology. In a short time, Tall Timbers became the premier outlet for research on fire ecology not only in the coastal plain, but the world.[14]

While Stoddard-Neel is not a formulaic approach to forest management, it does rest on several fundamental commitments. One is to single-tree selection. Single-tree selection was not a new technique in the forestry profession when Herbert Stoddard first started practicing it; its merits had been debated for some time.[15] But in Stoddard and Neel's iteration, the central proposition was a conservative marking strategy and the long-term survival of the forest, rather than the most efficient production of timber resources. Each tree taken, in effect, mimicked natural, small-scale disturbance events such as lightning strikes, blow-downs, insect damage, and other natural mortality. Such events open gaps in the forest canopy, allowing more light to hit the forest floor and making room for longleaf regeneration to take place. To open these gaps, Stoddard and Neel first selected the weakest trees for removal, leaving the stronger trees to build up the timber volume so that fewer, but more valuable, trees could be removed the next time around.

Unlike most foresters, Stoddard and Neel never predetermined the volume of any given cut based on total basal area calculations. They were less interested in taking an abstract growth increment from the forest than they were in attending to the details of which trees the forest could stand to lose and which it could not. Nor did they use diameter at breast height (dbh), a method for estimating the age and volume of a given tree, as a marking guide. They only marked an individual tree after seriously considering the specific ecological effects of removing it. Moreover, the marking of individual trees allowed them to leave as little evidence of their activity as possible. One of the end goals of the Stoddard-Neel Approach, in fact, has been to make harvesting operations largely invisible soon after their completion.

Along with single-tree selection and the conservative marking of timber, the other fundamental commitment of the Stoddard-Neel Approach is to fire management. As we have already suggested, fire is the lifeblood of the longleaf pine–grassland system. Without it, the forest would not only become something different ecologically, but it would also be more susceptible to catastrophic fires. The rough in the South grows quickly, particularly on good soils like those of the Red Hills, such that forest fuel levels can reach a critical mass after just a few years of fire exclusion. Under such circumstances, a simple spark can quickly escalate into an inferno. In fact, as we were in the midst of working on this book, the consequences of fire exclusion became clear in the summer of 2007 when major fires consumed several hundred thousand acres of timberland to the east of Thomasville around the Okefenokee Swamp. The Stoddard-Neel Approach eliminates such a threat by applying frequent fire on a rotation of at least every two years. In the process, frequent fire creates ideal growing conditions for longleaf pine timber and all of the ecological richness that accompanies it.

Stoddard and Neel also developed several innovative practices in their dealings with loggers and timber buyers. The simplest was their standard logging contract. As Stoddard wrote in an early memo to potential buyers, "the requirements governing the cutting of timber are so different on the properties in question from ordinary commercial timbering operations" that all parties should be prepared to be exceedingly careful while working in the woods.[16] Among several other provisions, loggers were not to damage any unmarked timber, create any new roads, or leave any refuse. And Stoddard and Neel were always on the ground to prevent any departure from the contract. To an industry accustomed to cutting most anything in view and leaving a mess in their wake, these terms may have seemed overzealous, but Stoddard and Neel knew that the hunting preserves contained some of the best timber remaining in the coastal plain, and sawmills would do whatever was necessary to get at it—including adhering to a set of strict rules for how to conduct a careful harvest. Close supervision and control of logging activity in the woods quickly became a hallmark of the Stoddard-Neel Approach.

What is perhaps most impressive about the Stoddard-Neel Approach has been its ability to sustain a multiaged forest through single-tree selection and prescribed fire during a time when the southern forestry profession as a whole moved toward the even-aged forestry, clear-cutting, and fire exclusion demanded by the pulp and paper industry. By the time Leon Neel began working with Herbert Stoddard in 1950, pulp and paper forestry already dominated both the professional literature and much of the coastal plain landscape itself. The pulp and paper market moved into the South in the 1920s and 1930s with the arrival of the first kraft paper mills, and it exploded after chemists such as Georgia's Charles Herty refined the processes to produce newsprint and white paper from young southern pines. By the 1950s one of the standard forestry methods in the South was short rotation, even-aged management, based on a plant, clear-cut, and replant model. And longleaf pine trees were not the species of choice under this system. Because longleaf grew slowly in the developing stages, the industrial and governmental forestry interests focused nearly all of their research and development on the faster growing loblolly and slash pine.[17]

As the coastal plain outside of the Red Hills grew up in a vast monoculture of even-aged plantation pine, the Stoddard-Neel Approach continued to center on the longleaf (and to a lesser extent loblolly) remnants, aiming its timber cutting at very different segments of the forest products market. They grew and harvested trees almost exclusively for the sawtimber and pole markets, which both required older timber and brought higher returns. Stoddard and Neel engaged with the pulpwood market, but only to sell the by-products of a sawtimber or pole harvest, or to conduct necessary thinnings in the process of forest restoration. They used the pulpwood market to fully utilize a timber harvest, in other words, but they never managed forests specifically for pulpwood.

The southern pulp and paper economy has contracted in recent years, and the southern forestry profession is in flux as a result. Today, one is just as likely to hear a coastal plain forester recommend planting longleaf for sawtimber production as loblolly for pulpwood. Government initiatives such as the Conservation Reserve Program even encourage farmers to plant longleaf and native grasses in fields no longer used for commercial agriculture. The opportunity exists, then, for extensive restoration of longleaf pine across the coastal plain. Moreover, some knowledge of fire hangs on in most areas of the rural coastal plain, and the serious issues of smoke management notwithstanding, many residents relish the thought of resurrecting a fully developed fire culture in the region.

There also has been growing public interest in understanding, protecting, and restoring the historic longleaf pine forests of the region. In fact, the longleaf ecosystem has received a tremendous amount of attention in the past decade or so. Nature writer and activist Janisse Ray has done as much as anyone to call the public's attention to the longleaf forest. Her 1999 book, *Ecology of a Cracker Childhood*, and her 2004 follow-up, *Wildcard Quilt*, are gorgeous expositions on living in the ruins of the longleaf pine's former range and appreciating the few remaining refugia.[18] Others are working hard on repopulating that range with its former inhabitant. The Longleaf Alliance, a nonprofit group of forest scientists based at Auburn University, is working on making longleaf a viable reforestation alternative. The Nature Conservancy, as well, continues to identify remaining tracts of longleaf and works to prevent their destruction. In addition, the land managers on several military bases in the coastal plain take seriously the management of their longleaf forests, and many state and federal forestry and wildlife divisions are eager to protect their remaining forests and help to restore those on private lands. There is, in short, much to be hopeful about if one values longleaf woodlands.[19]

But one must not mistake the increasing numbers of planted longleaf pine for ecosystem restoration as the Stoddard-Neel Approach would define it. In most systems of forest management, there is a termination point when it is time to start the forest over. It may well be that, once they reach commercial maturity and the market is right, these planted longleaf pines will be clear-cut and replanted, mimicking the industrial agronomy of the timber plantation. Even growing on a long rotation for older sawtimber still implies having a starting and an ending point. The Stoddard-Neel Approach does not work that way. The terminology of commercial forestry is largely absent in Stoddard-Neel, because there is no beginning or end to the forest, only a continuum that reaches into the past and stretches far into the future. Some of Leon Neel's favorite expressions—"we never terminate a forest," or "we practice forestry beyond one generation"—vividly illustrate the point. What we do with our environment in the present, he believes, should always be done in reference to an ecological timescale far larger than we are used to considering. There is unprecedented interest in the longleaf pine–grassland system at this moment, and this book is Leon Neel's argument that the Stoddard-Neel Approach should play a central role in its protection and resurrection.

This memoir is one result of about five years' worth of extensive collaboration between Leon Neel and the two of us. As we have already suggested, the idea for this book began at the Joseph W. Jones Ecological Research Center at Ichauway, a former quail preserve south of Albany, Georgia. Ichauway has its own interesting story. Robert Woodruff, the longtime chairman of the Coca-Cola Company, cobbled it together in the 1920s and 1930s out of former agricultural and turpentine land. Its 29,000-acre land base makes it one of the largest contiguous tracts of privately owned land in the state. When Woodruff died in 1985 the board of his foundation knew to do something special with the place. Joseph Jones, Woodruff's closest assistant and confidant for many years, took the lead and consulted with the necessary experts. They decided to make it an ecological research station, and incorporated it as such in 1991. Lindsay Boring, a forest ecologist at the University of Georgia, came on board as the center's new director, and he set out to hire a full staff of accomplished scientists interested in the longleaf-grassland system. He hired conservation biologists, wildlife biologists, forest ecologists, plant ecologists, herpetologists, hydrologists, limnologists, land managers, and educators. In addition, the steering committee appointed a scientific advisory committee that, over the years, has been stacked with such ecological luminaries as Eugene Odum, Gene Likens, Mac Hunter, and Jerry Franklin. Within a very few years, the Jones Center at Ichauway became a home base for serious scientific research on the longleaf pine biome.

Even so, they felt they lacked the historical depth of field that only a lifetime spent in a particular place can produce. Enter Leon Neel, who became a member of the Jones Center's scientific advisory committee in 1998 and time after time offered counsel that only a lifelong land manager could provide. In the process, the Jones Center became committed to studying and promoting the Stoddard-Neel Approach. It was not long before several staff members realized that Leon's story—and his stories—needed to see print. Lindsay Boring, Kevin McIntyre, scientists Steve Jack and Bob Mitchell, and Leon himself began the search for someone to help them out, which is how we ended up in that parking lot with Leon Neel. We want to thank Lindsay, Kevin, and all of the other people at the Jones Center who supported this work, both intellectually and financially, and became friends in the process.

We began this project by conducting a series of oral history interviews in the summer of 2004. The first finished product of that process—an extensive oral history transcript of interviews with Leon and several of his most important colleagues and collaborators—is now housed at both the Jones Center and the Forest History Society in Durham, North Carolina.[20] We would like to thank the Forest History Society for their support during the completion of this project, and for allowing us to reprint portions of an article previously published in *Forest History Today*.[21] Both the Jones Center and the Forest History Society hold copies of the original tapes. Anyone interested in the full range of our recorded conversations with Leon should consult those transcripts—or, better yet, listen to the tapes.

Not long after we finished that project, and as a result of continuing support from the Jones Center, we began the process of winnowing and refining the rough material of that transcript into a polished memoir. We initially hoped that, through the judicious use of the cut-and-paste function, we could pull something publishable from the fat oral history transcript relatively easily. But our process turned out to be more involved than that. As we aggressively edited and moved material around, we had to provide a substantial amount of connecting content as well. We developed rough drafts of each chapter and filled in some gaps in the story the best we could. Then we shared the results with Leon and paid him a series of visits to work through the drafts thoroughly. Those conversations were not recorded (except indirectly within these covers), but they were very much a continuation of the initial interview process. During those conversations, Leon provided additional material, reflected herein, that does not appear in the original oral history transcripts. After working through each chapter this way, and ghostwriting some transitional sections that linked materials from the oral history, we have a finished product that is the result of a truly collaborative process. While not every word is Leon's, he actively participated in the editing and rewriting process and approved every change we made. Although we have rendered this book in memoir format, it is the embodiment of our conversations.

Putting together this book was a cumbersome process, and we often grew weary from the amount of time and energy it took to pull off (and no doubt we tried the patience of both Leon and our supporters at the Jones Center). But the great side benefit was all the time we have been able to spend with Leon and his wife, Julie. We first met Julie that same day we met Leon and saw Greenwood; she joined us for lunch after our morning tour. Leon and Julie have been partners in their lives and work for sixty years, and they are almost always together. Julie grew up on a dairy farm in Sumter County, Georgia, just outside of Americus, and she knew well the patterns of rural life in the South when she met Leon at the University of Georgia in the late 1940s. They married in 1948 and together moved to the Red Hills shortly after graduation. They have been living there ever since. Like Leon, Julie is an accomplished naturalist, and her positive identifications of plants and animals from her and Leon's past, usually with scientific names included, lend essential detail to this book. She is also a master butterfly gardener, and getting to watch Julie's butterfly garden change through the seasons was alone worth our many trips to the Red Hills. Although you rarely hear Julie's voice directly in this book, her influence is always present. She was often in the room during our work sessions—or was called in by Leon—to offer clarifying details about people, places, and events. She was at the center of our collaboration. We cannot thank Leon and Julie enough for sharing themselves and their home with us in the process of this project.

The narrative that follows is technical where it needs to be, but we have written and edited with the informed lay reader in mind. Leon's way of managing land is not overly technical, so there is little need to get mired in the minutiae of forestry science and economics. Moreover, the scientists and land managers at the Jones Center have been doing important work to fill out the scientific dimensions of Leon's management practices, so readers interested in that sort of an approach should consult the array of publications that they have produced on the subject.[22] We have tried to dwell instead on the philosophical and practical dimensions of the Stoddard-Neel Approach and the years of experience in the woods and fields of the Red Hills that inform it. And we have woven in as well the many stories that make up the history of the Stoddard-Neel Approach and that function as parables for its practice.

The memoir that follows is divided into four chapters. The first focuses on Leon's early years in the Red Hills. To be frank, Leon has never been entirely comfortable with this section. He often asked us what any of it had to do with his land management philosophy and techniques. But we insisted that his childhood explained a lot about his passion for the land and the sorts of commitments reflected in his management. The second chapter focuses on Leon's years of work with Herbert Stoddard, and his sense of the importance of Stoddard's pioneering research and land management. Chapter 3 looks specifically at the formation and early years of Tall Timbers Research Station, to which both Stoddard and Neel contributed in important ways. The final chapter, and in many ways the most important one, outlines the Stoddard-Neel Approach in considerable detail. The book ends with a brief afterword by Jerry Franklin, a distinguished professor in the College of Forest Resources at the University of Washington, who is one of the leading contemporary students of and advocates for ecological forestry. As he makes clear, he is also an admirer of Leon's work.

The result, we hope, is a book that will be useful to foresters and forestry students, landowners and land managers, environmental historians, and concerned citizens alike, one that communicates the sense of urgency that was never far below the surface of our conversations. For as much as Leon Neel has done over the past half century to protect and nurture the spectacular longleaf woodlands of the Red Hills and other parts of the coastal plain, and to build on the legacy of Herbert Stoddard, he has also seen too much of his careful work disappear in the blink of an eye, as landownership changed hands and the commitment to conservation waned, as financial exigencies required (or greed encouraged) a landowner to cut too much timber, or as the desire for quail maximization led some to thin their timber in ways that damaged the ecological integrity of the forest-grassland environment. It takes four hundred years to grow a four-hundred-year-old tree, Leon often reminded us, but only a few minutes to cut one down. In that sense, the most challenging aspect of the Stoddard-Neel Approach might be the restraint it requires of the landowner and land manager. Let this be a call for such restraint.

CHAPTER 1

Growing Up in the Woods

One of the most important tenets of the Stoddard-Neel Approach is a deep appreciation of the woods that one is managing, an appreciation born of intimate experience working and being in the woods. While the approach itself is the product of my experience working with Herbert Stoddard and of professionally managing the woodlands of the Red Hills over more than half a century, much of my appreciation for this landscape and its history was a product of my childhood in the region. I grew up in this landscape, and that has forever marked it as significant for me in ways that extend well beyond my career as a forester and land manager. Growing up in the woods provided the foundation for my land ethic.

My family's roots run deep in Thomas County, Georgia. My immediate family—my mother, father, brother, and I—lived with my grandfather in a house that he built around 1900 on South Broad Street in Thomasville, Georgia, and the house is still there. It is called the Neel House, and today it is an office building. It is a fine house, and I am glad somebody owns it who can afford to keep it up.

Even though we lived in town, I grew up during the Great Depression on the land and in the woods, which was important in terms of the development of my land management techniques and land ethic. We depended on the farm for our livelihood, and particularly for much of what we ate, so we had to pay close attention to it. I also count myself as fortunate for having been born into a family that had respect for the land, even though it was not precisely formulated or expressed as a land ethic. My parents and other family members appreciated the land, and they recognized that what they enjoyed and what sustained them was coming from the land itself. In that respect, they were not unlike many rural people at the time, though that sort of connection often has been lost in this modern world.

I grew up in a house full of people. My brother, Howell, was seven years older than me, and he was my only sibling, but there were four families living in my grandfather's house. My grandparents, of course, lived there, and three of their children's families were in different sections of the house. My grandfather actually purchased the whole block way back before he built the house, and he divided it into three big lots. His house was closest to town, facing Broad Street. The next house belonged to a daughter, and the next house belonged to a son. He had nine children total, so he had two of them next door and three under his roof. There was a lot of coming and going in the early days. It was a very active place. We all gradually moved out of the house; my family and one aunt were the last to live there. In the late 1930s, when I was about twelve years old, my father built a house right next door, in what used to be the barnyard. Growing up in town back then was not so different from growing up in the country.

As I remember it, Thomasville was a wonderful place to be a child. It was a small community; you knew just about everybody in town, and everybody knew you. Nobody ever locked their doors. The whole block was open territory for me. We had wildlife all around, including some quail. There was a park out front called Paradise Park, just across the street from our house. It was a beautiful park. It was heavily wooded with longleaf pine and had an open understory—you could see from one end of the park to the other. There was a grandstand in the middle with some paths leading to it, but that was the only development. Otherwise, it was a representative piece of the longleaf belt. And I remember vividly that there was still wiregrass out there. It was not healthy-looking wiregrass because of human use, and it was mowed occasionally, but it was wiregrass still. There was not a solid stand of it, but there were clumps here and there. I went back to look for it a couple of years ago, but they have been closely mowing it for so long that I could not find anything. It would be an interesting ecological experiment to put a fence around that park and start putting a fire regime through it to see what would happen. You might just see some of the former understory vegetation come back. Fire is amazing that way.

I was the youngest child, and my father was the youngest child, so I had many older cousins. In fact, several of my oldest first cousins were older than my father. Some of them did not live in Thomasville, but I still had more people looking after me than I ever knew about. Even though I might have thought nobody was watching me when I was squirrel hunting around the block, they kept a pretty close eye on me—my aunts and uncles and cousins. There were times when I thought I was off by myself in the big, deep jungle, but in retrospect, I imagine that I was rarely out of the sight of someone in my family.

While I grew up in town, I was from a farming family. That was a pretty common situation back then. My grandfather Elijah Leonidas Neel was a farmer. He joined the Confederate Army when he turned seventeen, in early 1864, during the latter part of the Civil War. He was with the Fifth Georgia Cavalry in Savannah and saw action around Atlanta, up in South Carolina, and in North Carolina. He mustered out in North Carolina and had to walk home. When he came back from the war, the only thing to do in the Red Hills was to farm, so that was what he did. He either inherited some land or bought it somehow. But by the time I came along, he owned a few thousand acres of farmland, scattered around in Jefferson County, Florida, and Thomas County, Georgia. There was the home place, which was sold to wealthy northerners later on and became Sedgefield Plantation. I worked there for many years as their forester. Then there was the land where Julie and I live now. People of Scots-Irish descent settled this area and called this particular military district Glasgow, so my ancestors used that name for their land as well. My grandfather married my grandmother Mattie Heard, and through her they acquired the Heard Place, just northeast of Thomasville. At that time there were about twelve hundred to fifteen hundred acres out there.

My father had an indirect but strong influence on my approach to land management. He did a lot of things to make a living from the land, but I remember being particularly struck by the way he took trees from the woods. He had a little portable sawmill that he cranked up on occasion, and when I was young, I watched him move that around and cut a lot of trees on our land. He did not clear-cut, because there was no need to, but neither had he developed a true technique of removing trees for the long-term benefit of the forest. His approach was more intuitive, but it was effective from a conservation standpoint. I was not aware of a studied effort on his part to select particular trees for cutting, but he was making what we would today recognize as a selection cut, choosing certain trees for their value and leaving others there to continue growing. So in many ways, my land ethic sprang from watching my father make a portion of his living from the woods while also treating it with respect. He was able to cut a lot of timber without destroying the forest, and I wondered about that a lot when I was a little boy.

A lot of what I learned about the land growing up came in the context of agricultural production. My father looked after a scattered group of family properties. He had a degree from Georgia Tech and was an engineer for the state for a few years, but then he decided to start farming his father's land. So all I ever knew him to do was farm. He farmed most of my grandfather's land early on, and he had some cattle on one of the places. He also had some hogs and a turpentine operation that he started from scratch. He even had that small sawmill for a while. In other words, he spent most of his time in the woods and on the land. That is how I got indoctrinated as a woodsman. And my mother was just as much of an outdoor person as my father. She loved to fish and to hunt as well. She was a particularly good turkey hunter. So I was always either going hunting or going fishing, or doing something outdoors with one or both of my parents.

There were tenant farmers on many of the properties that my father farmed. The land had little market value during the Depression, so it was really a subsistence situation for all of us. The people on the farms who did the work were both white and black, though there were more blacks than whites. I am sure there was a lot of dissatisfaction with the system, but in those days most everybody had to live off of the land, my family included. I never considered myself to be growing up in a rich family, but we were well-off compared to many of our neighbors. We were fortunate to own land, and I continue to enjoy a small part of it today.

The land itself encompassed a whole spectrum of environments and land uses. Back then, there were twelve hundred acres on the Glasgow property where I now live, much more than we have today. Daddy ended up buying this property when my grandfather's estate was liquidated in 1939 or 1940, which was the best thing for me that he ever did. I do not mean that from an economic standpoint, although the land has become economically more valuable. Being able to live on this land has been important to us, and so it was the most meaningful gift he could ever have given to me when he acquired this land.

The land on which we live is almost all forested today, but it was mostly cleared and farmed when I was a boy. I know of at least eight wells on the place, and there would have been a minimum of one tenant house at each well. Sometimes you had two or three families pulling out of the same well. So this land was farmed extensively, and there were a lot of people living on it. Even the bottomlands were farmed, which required a lot of work. The soils were very good in those bottomlands, but the water was difficult to control. The land had to be ditched to control the runoff, and whoever laid those ditches out had to have a good understanding of bottomland hydrology. It is amazing how much they could do with so little in those days. This whole property is good red clay soil, but it is hilly and prone to erosion, which makes it tough land to farm in a conservation way. Nonetheless, it was farmed extensively, as was much of the Red Hills landscape, which means it looked different back then than it does today.

Despite all the agricultural activity, we had some beautiful longleaf land in the 1930s. The Heard Place had a few hundred acres of longleaf that, in my memory, looked just like Greenwood Plantation does today. It was a mixed longleaf-wiregrass stand with a multiaged class of timber. I would not have described it that way as a child, of course, but I knew it was fantastic. Daddy used to hunt quail there a lot, following one of his dogs. I shot my first quail with him there, and I still remember the exact spot where it happened. There is a little depression pond in the middle of those beautiful pinewoods, and there was good cover around it. The dogs pointed, we flushed the covey, and one swung on my side. I had one clear shot by myself, and I swung over and killed that bird. It was a cock bird. I can find the spot where he fell right now. Moments like that really connect you to the land. Sadly, this area today is an improved cattle pasture without a tree on it.

When I was young, the portion of the Glasgow land that we live on today had one white family and three or four black families living on it. The number of wells on the place indicates more families earlier on, but when I came up, people were slowly moving off the land. Those who remained all made their living from the land. Hogs were a staple, for instance, and we always had hog killings. The families would get together to kill a hog or two when the weather was right, and then we would smoke our own meat. Hog killing was a great time. Hogs were killed in cool weather, because pork spoils so easily. The colder it was, the better it was for hog killing. But a lot of times, the stored food would run out early, and we would have to kill hogs before it got to be the dead of winter. Hog killing was a full-day's process, and everybody had a job—the men folk, the women folk, everybody. The process got started early in the morning. Daddy had a little .22 rifle, and he usually shot the hog between the eyes. Then we processed it right then and there. We had a big syrup kettle, and for hog killing time we would fill it with water and build a fire under it to get it boiling. Then we put the hog in the kettle, which scalded it and made it possible to get the hair off without any trouble. Then we butchered the hog. It is true what they say: every part of the hog was utilized, everything but the squeal.

Hog killing was hard work, but it was also a great social occasion. To a youngster, though, nothing beat cane grinding. We planted a little sugarcane, as did most families in the area. This was not the sort of commercial operation many are familiar with today. Cane syrup was important to us in the household as a sweetener, but we never sold any syrup. To grind the cane, we fed the cane stalks into the mill, which was turned by a mule attached to a crooked beam over the mill. He just went around and around in circles, grinding cane. When the stalk came out, it was crushed completely and about 90 percent devoid of juice. The pumice—the remains of the milled stalks—was discarded on what we called the pummy pile, where the kids used to play. The worst thing about that was the yellow jackets—they came around for the sugar, so I got stung a lot. But it was worth it, of course, because we got to suck on the cane and drink the cane juice. There was always a fire, and the best cane grindings went long into the evening.

Everyone also grew sweet potatoes, which were a food staple. After harvest, folks banked the potatoes to make them last. They would clear a piece of well-drained ground about eight feet in diameter, and then they would lay a deep mat of longleaf pine straw down. After they dug up the sweet potatoes, they stacked them up on the pine straw bed. Folks would use slash pine or loblolly if they had to, but longleaf was the most desirable straw because it made the most durable mat. That is one of the little examples of how I began to understand that the longleaf pine was a special tree. Once they got the pile of potatoes in a conical shape, they covered them with another layer of pine straw and made a teepee-like structure around the pile with boards, tin, or most anything available that would shed water. They covered that structure with dirt and left a little hole at the top for air to circulate. They also had one board at the bottom that they could slide aside to reach in and get some potatoes. That was a sweet potato bank, and the potatoes lasted all winter that way if they were not eaten. The bank kept them dry and cool so they would not rot. You could not just go up to anybody's potato bank and start pulling potatoes out of there, though. You had to get permission. We grew the sweet potatoes mostly for household use, but Daddy sold a few when there was a market for them. Most farmers planted them for subsistence, but if it was something they could plant a little bit more of to sell, they would. That was the way it worked with most household crops.

We would get sweet potatoes to take night hunting. The folks on the place always had an old dog or two that had some hound in them. Some of them might have been well bred, but most of them had a little cur in them as well. We would go possum hunting a lot, which simply meant igniting a torch and letting the dogs chase whatever they could find. On cold nights, which were the most fun, we always stuck some sweet potatoes in our pockets before we left. We would get out in the woods a mile or two away to let the dogs go run up something, and if it was cold enough we would stop and build a little fire. We would let it burn for a while to develop a few coals, and then make a little hole under the edge of the fire and put those potatoes in there. Those sweet potatoes were the perfect size, I thought. They were dark red in color, and when you baked them in the fire, thick syrup would come out of them. That was a real treat. It was more fun to go for the sweet potatoes than it was to catch a possum.

Daddy planted several other crops. He grew corn for both food and feed. In the spring we picked some for the table—what we called roasting ears or green corn. You could not beat green corn in the spring of the year. Most of our corn was planted for cattle, hog, and mule feed, though. I do not remember Daddy ever selling any corn, but he may have sold a little here and there. But it would not be like selling corn today. You would sell some if your neighbor needed a hundred bushels, and if we needed a hundred bushels of corn we could buy it from the neighbors if they had some to spare.

Some of the corn we grew was made into hominy by treating it with lye. Most folks in the countryside made it. I liked hominy, but my mother did not have much trust in the process, so we did not have it often. But I clearly remember tenants making it on the farm. They dried the corn and removed the kernels from the cob. They would then boil the corn in a large pot with pure lye. The lye separated the husks from the kernels. They washed the corn several times after boiling to remove the excess husks and to clean off the lye. Then they simply boiled the tender corn and put a little salt and butter on it. It was good, but that lye made my mother nervous.

I also remember cotton being grown throughout the Red Hills, but it was not a major crop on our place. It had been the major cash crop in the region for many years, but it was in decline by the time I came along. Prices fluctuated so much, and then the New Deal programs like the Agricultural Adjustment Administration encouraged farmers to plow up their cotton. I think farmers around here were just tired of messing with it.

Around World War II, my father shifted from the old type of subsistence farming, because having access to money became more important just to live. I do not mean to imply that we needed a lot of money, but the fact was that you could live on very little cash before the war, or at least you could certainly survive. After the war, there was a systematic change in the way southerners farmed: it became more of a business. I think this change came earlier to other regions; economic modernization came a little late to us. My father went into several different kinds of farming to increase our market production. I guess you could say we were diversifying as well as commercializing.

We got into truck farming to bring in more income. We planted some truck crops like watermelons and okra, which was risky. In planting truck crops you were gambling on the market from here north all the way to New York. And it was hard work, too. Anyone who has worked an okra field without gloves would know what I mean. Okra has a prickly, hairy pod, and the leaves are prickly as well. We cut okra with a pocket knife, and we had to reach in through the leaves to find the right-sized pod to cut. It could eat you up. On top of that, we picked in the dead heat of summertime, so we were sweating, and that okra plant had you breaking out in rashes and itching all over. Okra brought a lot of money if the market was right, but it was such a labor intensive crop that you could not plant a lot of it. We would plant about a half acre, which was a lot of okra. I remember the first hamper of okra I ever picked. We carried it to market, and that hamper brought eight dollars, which was an unheard-of price. Before that it was down to two, three, or four dollars, but all of a sudden there was a demand for it. But the market might bottom out by the time of the next sale, so it was always a risky business.

The same was true of watermelons. They had to be loaded into cars on the railroad sidings, and there were several railroad sidings in the area to which we could carry our watermelons. We sold them to what was called a jobber. He bought them here and then turned around and sold them somewhere else. There were always one or two men in Thomasville who handled all the watermelons; they made the arrangements for the boxcars to come in and pick them up. Whenever the watermelon boxcars arrived, all the local farmers brought in their melons to sell. All the farmer had to worry about, in terms of transportation, was getting them to the railroad siding. Growing watermelons was still a small, family-oriented business, not the big business that it is in some places today.

We also expanded the cattle business after the war. In earlier days, my father grazed cattle in the woodlands, like most southerners had done for a long time. But by the 1940s we converted some old fields into improved pasture. The state was making a push to make cattle a viable cash alternative to cotton, so improved pasture was on the increase. They were also importing improved cattle stock and trying to strengthen the markets. But these were all experimental measures, and the farm economy was far from stable. Stable farming is an oxymoron anyway: nothing is stable in any type farming. We were all figuring it out as we went along.

Many farmers in the area also started planting pine trees in their former cropland around World War II. I have some slash pine out in the front of my house that my father planted in 1941, and I was right there with him. We planted them with the help of three or four men. In those days we did not have a dibble, which is a simple tool designed to plant trees, so we planted the seedlings with grubbing hoes. There were several reasons for planting former agricultural land to pine trees. First, there was a growing market for pulpwood. More importantly, I think, small patch farmers could not make a living anymore, and throughout the region many of them were heading to the towns and cities, or up north. In my part of the South, southern Thomas County, you could not have a fifty-acre field if you wanted one, because the terrain was so rolling and uneven. So when farming became mechanized, we could not expand production in the Red Hills to the extent that they could in flatter areas of the coastal plain, not to mention the Midwest and the whole prairie area. Such large-scale production farming was killing the little patch farmer. We could not compete with them, especially in corn. They grew so much of it in the Midwest that their transportation rates were much cheaper. On top of that, the big seed companies had written us off. They were selling the best seed for cheap to the biggest farmers. As I discuss in more detail later, Thomasville's Greenwood Seed Company stepped up in the 1950s to develop a viable hybrid corn seed for southern farmers, but most small producers were already converting their agricultural land by that time. So, with the encouragement of both timber companies and conservation agencies, more and more people started planting pine trees in their old fields, and in some cases the pines just came in on their own.

We planted pines in our big fields, because they had been badly washed out. I think that was the case through much of the region. Daddy was a forward-looking man, and he always liked trees anyway. He knew what would happen if you just abandoned a field and did not do anything with it. A field that is already eroding will continue to do so until some succession plants seed in, which can be a long time in some cases. So he went ahead and planted trees to stabilize the erosion. He did not live long enough to see these trees get big, though we might have thinned them one time before he died. He did not plant those trees just for money, that is for sure. There was a conservation purpose to the planting of them. There was no straight conservation line to follow back then, but my father did a lot of things because they were the right things to do for the land.

Erosion was a big problem during the 1930s. The Red Hills soils, of course, are high-quality red clay soils, but the problem with agriculture here is the terrain. The region is called the Red Hills for a reason: if you clear all the vegetation off of a hill and farm it, you are going to get some kind of erosion. Some of the soils eroded more quickly and severely than others, depending on the field sizes, the contours, the slopes, and the soil types. Tall Timbers, for instance, has some erosion ditches that you could hide a pickup truck in. If you put vegetation back on the land you can eventually stabilize such erosion; you will not get rid of the ditch, but you can stabilize the erosion. Of course, the topsoil had long since gone from a lot of the areas that had been farmed, so that changed the types of vegetative communities that grew back there with the trees. Growing pine trees, then, was a sensible move for farmers such as my father.

Before World War II there was still a fair amount of agriculture in Thomas County, but not necessarily in the Red Hills section. North of the Ochlocknee River there is good farm land, and over on the eastern end of the county there are good flat lands with different soils that hold more water. The northern edge of the Red Hills really stops at Thomasville. What we call the Red Hills today is generally defined by a series of boundaries: the Aucilla River to the east, the Ochlocknee River to the west, and the Cody escarpment to the South, which is where the Red Hills drop into the flatwoods south of Tallahassee. The land immediately outside of the Red Hills remained heavily agricultural after World War II, but within the boundaries of the Red Hills proper, the landscape was becoming more forested.

I spent a lot of time hunting and fishing when I was growing up. My Daddy was a good hunter and a good shot. He hunted for the true sport of it, and for the meat as well. He did not hunt for glamour or trophies or anything like that. He sometimes shot ducks, for instance, on the big lakes in north Florida such as Lake Miccosukee, which were renowned for their duck hunting. But he also shot at Heard's Pond, which was on our property then. I have fond memories of being with him in the boat when he shot ducks there and I was still too little to shoot. My father also loved turkey hunting. We did not have a spring gobbler season like we do today; we had an autumn and winter turkey season that coincided with quail season. It always opened on November 20 and closed on the last day of February. The only thing my father did not shoot was deer, because, believe it or not, there were so few around. We did not have many deer in the Red Hills when I was growing up for several reasons. The screw worm came in around this time, which reduced deer numbers, and there was more cleared land than there is now by a great deal. But, perhaps most importantly,

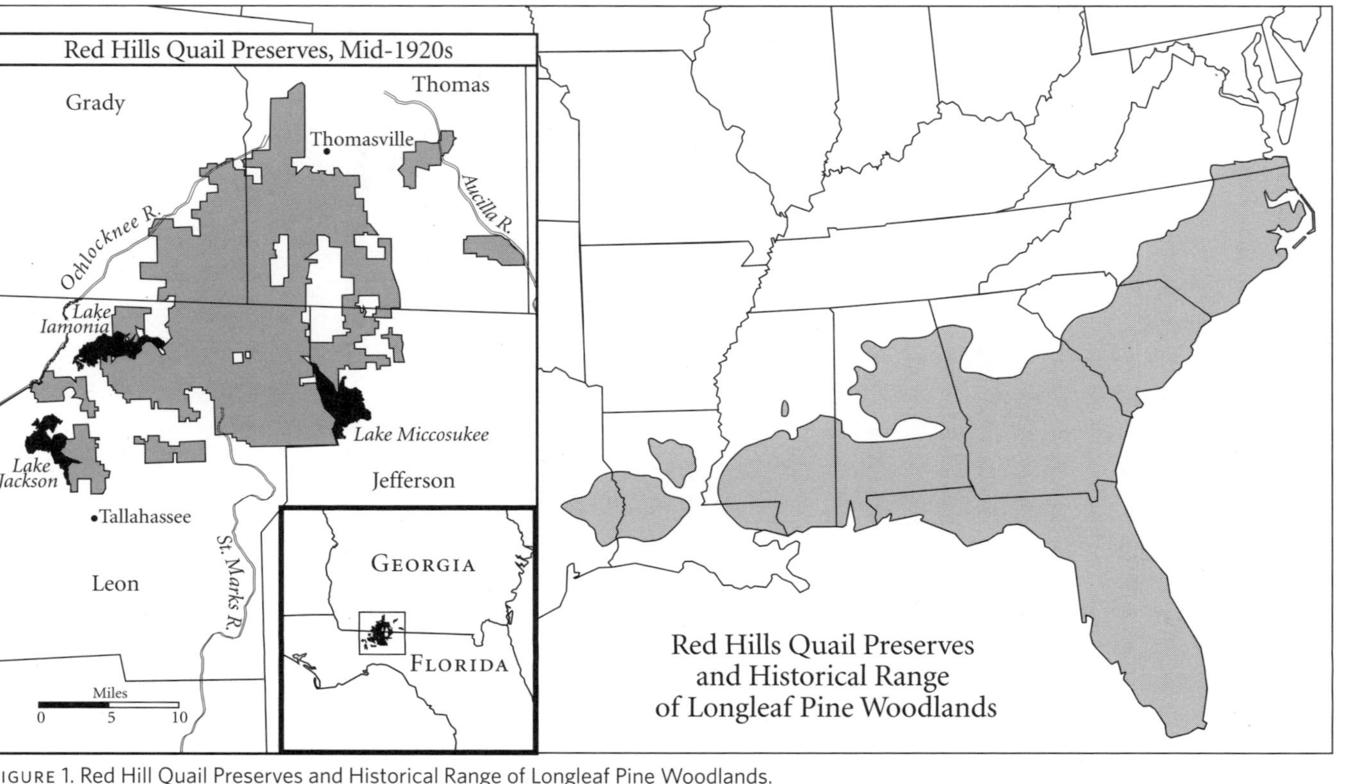

FIGURE 1. Red Hill Quail Preserves and Historical Range of Longleaf Pine Woodlands.

there were more people on the land who actively hunted, which quickly cut into deer numbers. So deer were scarce. As a hunter, Daddy took full advantage of what was available to him locally. He was not interested in going off to hunt pheasants, woodcock, or something that was not available locally.

My first hunting memory is of squirrel hunting. I was just a little fellow, and I used to like hunting squirrels. Daddy had a short, full-choked barrel .410 single shot. That was what I used, because I could not shoot a rifle well at that time. We would go out and ease through the woods. Daddy knew of some good areas for squirrels. These were grey squirrels, what we call cat squirrels to distinguish them from the region's distinctive fox squirrels. We never shot a fox squirrel. My father did not believe in that, and we still do not believe in it today. Fox squirrels are special animals to us. We have raised a lot of them and have come to know them, and they are gentle, caring, and beautiful creatures. That is a humanistic point of view, my preference for fox over cat squirrels, but sometimes I just cannot help that.

We mostly stalked the cat squirrels. Sometimes we would sit down by a tree, but usually we would move along through the woods. When I became old enough I used to shoot squirrels on that block of my grandfather's in Thomasville, too. I was about ten years old before I could hunt squirrels with my .22 rifle. I would get up before school, especially in the fall when pecans were getting ripe, and do a bit of hunting. My aunts would always want me to kill those squirrels, because they needed the pecans for cooking and baking. I would get up, make a course around the block, and kill one or two squirrels in the morning, maybe three if I was lucky. Other times I would not get any, but I always had a good time.

My father had dove fields as well, and we hunted them a good bit. Doves were, like much of the wildlife in the Red Hills, the products of particular farming practices. My father would plant a field to peanuts and then he would put hogs in there to fatten them up. That made a perfect dove field, because the hogs were eating the peanuts and wasting a lot, and the doves came in for what was left. They also loved the bare ground where the hogs rooted. When my father let me start shooting doves, it took me a while to kill one, because my little .410 shotgun had a full choke barrel. The shot pattern comes out very tight with a full choke, so it was about like shooting a rifle at those doves. That gun was fine for shooting squirrels, because they were stationary, but doves were flying. I think Daddy was trying to teach me how to be a better shot, but it sure was frustrating as a little boy. But the larger lesson I want to convey concerns the doves and the fields and the way we learned about nature and land use through the experience of hunting. This was multiple-use land. We raised crops and livestock on it, but it also supported good wildlife populations, and that lesson always stuck with me. Back then, farm land was also wild land.

I was about ten years old when I killed my first turkey. My father planted feed patches for turkeys, usually with chufa, which is a rush grass that produces an underground tuber that turkeys love. Daddy would build a blind on the ground, on the edge of the chufa patch, and Mother would hide in there. She was a good yelper, and even though turkeys would come to feed without yelping, she would yelp to attract them. My mother killed a lot of turkeys. She would kill a hen if they wanted something to eat, but she preferred to wait for a gobbler. She shot some of the biggest gobblers you ever saw. We hunted by locating the turkeys in the evening and letting them go to roost. Then I would scatter them off the roost right at dark. That would force them to move around and look for each other at first light. We came back the next morning before they started looking for each other, and we called them up with a box call that Daddy made. That is how I killed my first turkey. We had scattered some turkeys right down here in the swamp on Glasgow. I had my little .410, and we got into the woods long before daylight. My father yelped a time or two, and the turkeys started talking. Finally, a hen flew into a tremendous pine tree that was in range, and Daddy said, "All right, shoot her." He just wanted me to kill a turkey, you know. I drew down on that hen and shot. I do not think she even ruffled her feathers. He said,

"Shoot her again." I shot at her again, and she did the same thing. The .410 did not have enough power to even scare her off. Daddy had a 12 gauge, and he said, "Give me your gun." I handed it to him, and he laid it down and said, "Now take my gun and kill that turkey." And I did. But I had to pick myself up off the ground after shooting that 12 gauge.

I also shot some ducks, though I never hunted on the lakes where they shoot ring-neck ducks. Lake Iamonia and Lake Miccosukee were the classic ring-neck duck lakes in the area. That is where hunters would place decoys and blind up in boats. We mostly hunted wood ducks on the cypress ponds around here. Wood duck shooting was subsistence shooting. If Daddy was doing something and had his gun and a wood duck came by, he killed him because he wanted to eat him. It was not a formal duck shoot like the ones on the big lakes. We sometimes had enough ring-necks to shoot at Heard's Pond, but not often.

The most significant part of the Heard Place property to me was a big, man-made pond that my grandmother's ancestors had built, called Heard's Pond. We always said it was three hundred acres, but that included its wetland fringe, an ecotone around the edge. The pond is in bad shape today because the dam was cut a few years ago. The water level is low, and the edges are growing up thick with vegetation. They repaired it one time, but not very well, and it blew out again. The pond was a natural slash pine flat where my ancestors built a long, low dirt dam. The pond was shallow, though there were some deep holes in there. Some of the holes probably went ten or twelve feet deep, but most of it was five or six feet. There were several cypress domes scattered through the pond, some of them quite large. A cypress dome is a thick stand of cypress trees growing in a slightly deeper spot where water collects. Now, a slash pine flat is basically a wetland. Wetlands do not have a tabletop terrain: they have some higher ground, some lower ground. The difference in elevation might not be much, maybe just a foot or two. Before this particular area was dammed up, the cypress trees occurred only in the lower, wetter areas, and the rest was slash pine grassland. So when my family flooded it, the cypresses were already there in the wetter areas, and they adjusted to the water depth and got bigger and bigger, encroaching outward because of all that additional water. You would not believe the frog calls, and alligator grunts, and everything else we would hear out there at night. It was just a wonderful place.

We had two shelters out in one of the domes in the middle of the pond. The oldest shelter was built fairly high off the water. I would say it was close to ten feet off the water. The walkway going up to it was at an angle—there were no steps, it was just a steep ramp—and I remember being sort of scared of it when I was little. It looked like a long way down there to the water. The shelter was solid, just basically a floor and a tin roof, and the roof was supported by several columns. It was connected to another shelter that was only about four feet off the water. We called that the cook shelter, but I do not remember anybody cooking down there.

When I was little, Daddy built another shelter in the same area. I was with him when he first went in there and picked out where he wanted it. He went back and got a couple of men and a crosscut saw, and they had to cut eight or ten cypress trees down to create an opening big enough to build the shelter he wanted. He used the fresh cypress stumps as the foundation pillars. They were not, obviously, lined up exactly right, but he picked out an area where he got them close enough. He was a good carpenter and he could build anything. On the Stegall Place that my granddaddy owned, which is within the Thomasville city limits now, one of my uncles had gone into the dairy business back in the late 1800s, and there was an old wooden silo there built out of virgin cypress. Daddy built the shelter using that virgin cypress from the silo, because it would not rot in water for a long time. That silo had been abandoned for a long time, and Daddy, who knew his wood, would go over there and get some silo cypress whenever he wanted something special. The shelter was relatively square. He had the studs going up to hold a standard tin roof. He sealed it up with boards all the way around, about wainscoting high. And then, where the roof started, he had that sealed up, and in between he put screen wire, to make it a screen shelter. It had a couple of doors, and it was connected to the other platforms by walkways built between the cypress trees. Then he built a walkway that went to the edge of the cypress dome, to get to open water where there were no trees. That was our swimming hole.

We used to go out there to the Heard's Pond shelter and spend several days at a time. You had to have a boat to get there, and Daddy built his own boats. They were built out of cypress and were purposefully narrow to navigate through the trees. Mother, Daddy, my brother, and I would camp out there. It was a magical place for a youngster.

We also did a lot of rabbit hunting when I was young. Most of the rabbit hunting occurred when they burned off the fields in the winter or spring. That was a big thing. I had several cousins my brother's age, and they came out to rabbit hunt, along with any friends who wanted to come along. We would surround a field, set the fire, shoot the rabbits when they came out, and then cook them. It was quite the party.

When I was young I had a mentor, a black man who lived on the place, named Rich Murray. He was, out of necessity, probably the best woodsman I ever saw. Woodsmanship, if you boil it down, is basically observation combined with experience. You need to be a careful observer, but you also have to know what you are observing. Rich and the area's other black residents lived off the land more than we did, and Rich was a superb woodsman as a result. He could tell you just about everything that was going on in the woods, and he also took a great deal of pleasure and pride in knowing the woods. In particular, he could always tell where a rabbit was bedded down. We often walked together on a farm road with a field on one side and a briar patch or a hedgerow on the other side. That was perfect rabbit habitat. I had the gun, and he and I would just walk until he saw one. I, of course, would not have noticed anything, but he would say, "There's one right there." He looked for the eyes; the eyes are so distinct and different from everything else down there. That is what he was concentrating on, and he could spot bedded-down rabbits better than anyone I knew. He was amazing. I shot many a rabbit like that with Rich.

Rich was the one I hunted possum with most of the time as well, and I was always in the woods with him. I also used to go rattlesnake hunting with Rich. We do not hunt rattlesnakes anymore because we are concerned about their conservation, but we did not think much of it back then. I have had some close calls with rattlesnakes, but, luckily, I have never been bitten. Anyway, Rich could spot a rattlesnake, and if he found one, he wanted it killed. He was not afraid of rattlesnakes, but he respected them and knew what they could do. The funny thing about Rich is that he would stand there stoically when he came across rattlesnakes, but a coachwhip snake, which is harmless, terrified him. They terrified a lot of people in those days because of the folklore associated with them. It was called the coachwhip, Rich told me, because they would rise up, grab you, and whip you to death. You still sometimes hear that myth down here today.

I never believed the folklore about coachwhips, but I had one experience that made me realize why others might have. I was taught as a young boy to always be aware of snakes when I was in the woods. One day, I found a coachwhip about six feet long in the road where we rabbit hunted. Because Rich did not like coachwhips, I decided to shoot the snake. But the snake somehow sensed what I was up to and disappeared into a thick, even-aged briar patch about forty feet in diameter. Coachwhips are extraordinarily fast, and when they move quickly they make a humming sound. There was a plowed field on one side, and I was in the farm road on the other side, so I had no doubt that I would be able to kill that snake. I went into the briars fully expecting to drive the snake into the field or road, where I could dispatch him easily. But after a long and futile effort, during which I found no trace of that snake, I stopped beating through the briars to think about my next move. All of a sudden, I began to sense that I was not alone, and then I saw, about twenty feet away from me, the coachwhip's head extended a foot above the briar canopy. He was looking right at me. As I quickly moved to bring up the gun, the snake just as quickly disappeared into the briars again. So then it was a cat and mouse game, with me beating the briars only to look around and see the snake observing me again with his head and upper body rising above the briars. Even though I had a shotgun, I could never get a shot, because he was too quick. Finally, I wished him well and left the briar patch to him. As I reached the farm road, I turned for one more glance, and there he was watching my departure. He seemed to be smiling.

Of course, we did a lot of quail hunting when I was growing up as well. We did not hunt quail with a great deal of decorum like they did on the big quail preserves. I enjoyed that later on when I had a chance to participate, but when I was coming up, we just went out with a couple of dogs to see what we could find. Daddy kept a quail dog or two, and he usually kept them down here where I live today. We still had all of my grandfather's land back then, so we had several places to hunt. There was no quail management per se on this land, because quail were basically a by-product of our style of farming and management. There was plenty of patch farming for subsistence, and the woods were normally burned, so we had plenty of birds and we knew where they were. My father would hunt with friends or family members, usually no more than one at a time.

Back then you could find quail almost anywhere, but you were most likely to find them in certain habitats, like around a field where there was feed and cover, or in hedgerows. I learned that lesson when I went hunting without a dog to help find them. In that case it was important to know where they spent their time. Quail are inactive much of the time, when they are nooning. They often get in a little sassafras thicket that might be only twenty feet across, but with clean ground underneath it and an overstory of sassafras scrub. Or maybe you would find them nooning in a plum thicket. When I went hunting without a dog, I went straight to places like that. I could sometimes find seven or eight coveys without a dog. I paid attention, and I knew where the quail were, or where I was most likely to find them. It always helped if you had a friend on such a hunt, because two people could cover more ground when you were trying to find quail.

We also did a lot of fishing when I was a boy. Heard's Pond was wonderful pond fishing. It had largemouth bass, chain pickerel, bream, and all sorts of other fish. Mother did most of the fishing, while my father poled the boat. She usually fished for bass with a casting rod and reel. The rods were short—I am not sure if they even made long rods then—and they were made of steel. They tapered to the tip a little, but they were square. She caught some big fish with those rods.

I remember taking my mother fishing the day after my father died. My wife, Julie, and I lived on Mr. Stoddard's property then, and there was a nice pond there. That day, Julie suggested we invite Mother out to lunch, and I thought that was a good idea. She came out, and I asked her if she wanted to go fishing. She thought that would be nice, so I took her back to the pond and we got into the boat, and she fished while I paddled. She caught a nine-pound bass that day, the day after Daddy died. That was not the biggest one she ever caught, but it was pretty close. That made her feel good instantaneously, which tells you something about how much she liked fishing.

We fished for bass with artificial lures, but we used live bait for bream. Bream fishing with cane poles and live bait actually helped to shape some of my ecological appreciation of these woods. In searching for appropriate bait, we would look for dead pine trees, especially those that were broken off or blown over. At the right stage of decomposition you can pry up the bark, and it will come off easily in big sheets. Well, that is where we got our fish bait. Under that bark, we found all kinds of insect larvae. We were after what we called flatheads, which are white segmented larvae about two inches long. There were two kinds that were very similar, and they both made good fish bait. One was a *buprestidae* insect larva, and the other was from the *cerambycidae* family, one of the longhorn beetles. They were both wood-boring insects that attacked freshly dead trees, often from lightning strikes. They lay their eggs, and when they reach the stage before they pupate, they are perfect for fish bait. They last a long time in sawdust, and the fish love them.

Another common bait we found and used was the oakworm caterpillar. We called them oakworms. As with flatheads, we routinely found two kinds of oakworms. I always called one of them the "heart's horn" caterpillar, but a lot of people called them "hawthorn" worms. It made sense to me to call them heart's horn because that hard part of the face is shaped like a little heart, and it's called a horn simply because any appendage on either end of a worm is usually called a horn. They will feed on the leaves of a lot of different tree species, but we found them mostly on water oaks and willow oaks, and I have seen them on live oaks on occasion. We used to get them off of oak trees that had survived fires, little groves of four- to six-inch saplings where the ground is nothing but oak leaves. After the adult worms hatch and lay their eggs, they start eating. That is when you can hit the bottom of the tree with the butt of an axe, or most anything blunt and heavy, and they will fall off. You could hear them fall, and then you just went around and picked them up. But the trouble with oakworms is they have a defense mechanism. They eat oak leaves and exude a caustic potash secretion. In fact, a lot of people call them potash worms. They will literally blister you if you hold them too long. It does not hurt, but it will make you shed some skin. When you are fishing with them, you need to wash your hands off after you hook one.

It was in the process of learning to fish for bream that I began to get curious about all those creatures under the tree bark and in the leaves. I began to learn the life histories of some of these insects, and, whether I used them for fish bait or not, I became fascinated by the insect diversity in the woods, and the important roles that insects played in the forest ecology of the region.

We did a lot of fishing in the area's rivers as well as in the ponds. The Ochlocknee River was a good fishing river. It was a wading river with black water and a sandy bottom, and we fly-fished it. Of course, you have to get the water level right. When it rains it will muddy up, and you cannot fish then. When the time of the year is right and the water level is right, however, you can have a good time out there, catching bass and bream. We mostly fished it in spring and summer. It usually carries more water in the spring, but if you have a May without severe rains it clears up and gets down to where you can wade it, and that is when it is the best. The fall is good, too, but anytime it gets right in the summertime, chances are you will do well. I fly-fished there with my father a lot, and mother always fished from the bank. Later on, after Julie and I got married, the two of us fly-fished out there as well.

When I was little, my daddy used to do something we called "reading the newspaper." One of the first times I ever went with Daddy that I can remember, we were riding down some of those little two-rut roads and we got to a sandy place. Daddy stopped the car, and he said, "Come on, we're going to read the newspaper." So we got out and he walked up there and started looking around, spotting tracks, and he would tell me what they were. There would be bird tracks, mammal tracks, insect tracks. You could get all sorts of information from reading the fine print, so to speak. And it was fun. I do not understand how someone can walk the woods and not see anything. Even if you do not see the animals themselves, there are signs of them everywhere. Julie does most of the newspaper reading for us right now, because she walks every day. She walks our roads and then comes back and tells me the news. She came in not long ago and said she saw where a turkey hen walked down the road a little ways. We have not had many turkeys on the place lately, so that was good news. I read the newspaper wherever I go, because that tells me what I might not see otherwise. It is one of the most valuable skills that my father taught me. To a certain extent, as I worked in the woods as an adult, reading the newspaper helped me to judge the health of the forest.

I first learned about the role of fire in the longleaf system by following my father when he was burning the woods. He was an intuitive woodsman, not a scientist. He burned the woods when he felt like the woods needed burning, when conditions were right for him to burn, and when he had the time to burn. To a great extent, that is the way I burn today, although I have to put a little more concentration into it.

My father burned on all of the land that my grandfather owned, and that ranged all the way from old-growth longleaf to old-field loblolly. We had examples of most of the region's forest types on our lands. Experience taught my father when and how to burn. Lots of people in the region burned the woods, and they had been doing so for a long time. It was the right thing to do, as he would have put it. There was no visible harm, and he did not burn up anything that had any immediate value. The burning kept the forest accessible to human uses, whether it was hunting, turpentining, cutting timber, or simply walking through the woods. His rationale was not so much scientific as it was practical, even though he was doing things that scientists value today. It was just what southerners did to their land so they could get more benefit out of it.

I may have learned about fire from my father, but my first direct experience with burning land was of my own doing. I was little and living in my grandfather's house in Thomasville. My grandfather also owned a vacant lot next to the house that had been cultivated land at some point. It had grown up in broomsedge and briars by the time I was coming up. It was a neat place. There were a lot of things out there of interest to a small boy. Then the next lot beyond that was a developed lot owned by a man named Mr. Robinson, a nice man who owned a bookstore in Thomasville. He had a Victorian house, a well-kept yard, and a row of camellia bushes down the property line between the vacant lot and his house.

I wandered over to that vacant lot one Saturday morning, like I often did, just to mess around there. I was wearing short pants like I always did as a small boy. I started walking through there, and the briars were so bad and so thick, I thought, "Shoot, this thing needs burning off." That gives you a sense of how instinctive fire management was to us back then. So I went back to the house. The kitchen had a big wood-burning stove in it, and there was a box of big wooden matches up on top of the overhead warming oven. This was later in the morning and the stove had cooled down, so I dragged a chair up there to get some matches out of that big box. Then I went running back over to the lot, struck one of those matches, and stuck it in that broomsedge, with no particular consideration about wind or anything else. I was just set on getting rid of those briars, so I ran straight across there and set the fire. When I stuck that match in there, it exploded. I mean, whoooom! It was obviously dry. It scared me so badly that I just turned around and ran back to the house as fast as I could and hid somewhere. I do not remember where I hid, but I was scared. Somebody put it out, and somebody else found me. I got dressed down pretty good about that. It was not a controlled fire, but, in its own way, it was a prescribed fire, my first one.

As a child, I had a few experiences with the quail folks and their plantations. Over the previous several decades, wealthy northerners had come to the Red Hills around Thomasville and bought up large quail-hunting preserves. Anybody with a legitimate reason to visit one of the quail plantations could do so. You could not just go out there to hunt, of course, but a lot of locals conducted business with the owners. My father had contact with the manager of the Whitney place right next door to our land. And he also knew the manager at Norias Plantation, a man named Marvin Sasser. Mr. Sasser was one of the finest plantation managers that ever worked in this area. They grew a little wheat down there. I remember one time Daddy went down there and bought a sack of wheat from Mr. Sasser to carry to a gristmill and grind it up for flour. That kind of access was always available. The plantations were private, but not private to the extent that they did not have contact with the community through their people.

Only a very few locals were invited to hunt on the preserves in the early days, but there were a few. They were usually people of importance in the community. We were not among them. I gained access to these plantation properties only after I went to work for Mr. Stoddard and proved my ability and my sincerity about managing land. The owners back then knew what they were doing, and they knew how to judge people who worked on their land. Mr. Stoddard had already proven himself years before and, of course, because I was associated with him that gave me a leg up. I still had to prove myself, however.

I only began to get invited on formal quail hunts in the 1950s. I was already a hunter and recognized the uniqueness of a plantation quail hunt, which was pretty good in those days. All the trappings were there; it was first class all the way. I did not experience these hunts firsthand when I was growing up, but I imagine they were roughly the same. Formal quail hunts varied from place to place. Most owners used what they called the Thomasville shooting wagon, which evolved over a period of time from a horse-drawn buggy into a mule-drawn wagon that carried a seat for the driver, seats for the shooters, gun boxes, dog cages, and storage on top of the dog cages. The cages were compartmentalized for the dogs. The shooters would ride through the woods in the wagon, and when the dogs pointed, they would dismount and get in position to shoot. Only two people were allowed to shoot at one time.

You had the wagon and the driver, and a couple of mules. Most hunts involved one hunting wagon, though sometimes there were two. You had a main dog handler as well as a man who assisted with the dogs. He watered them when the handler wanted, and changed dogs when the handler wanted. Then there was another rider or two, commonly called scouts, who kept up with the dogs when they were hunting under the direction of the dog handler. These hunts were well organized and well conducted in most cases. Two shooters usually rode in the wagon, and often there would be two, three, or four people on horseback, even though most hunts were relatively small. Each plantation had its own dress code and stuck with it. Generally speaking, all the people involved in the hunt, including the shooters, wore white jackets so they were highly visible, though some hunts used red jackets. The horses were groomed and given what they called a plantation cut, which means that they were clipped close above the lower part of their bellies and the upper part of their legs. They did not clip the bellies or legs so as to protect them against briars.

The properties were divided into courses, and you typically hunted one course in one day. You would not go back to hunt that course for at least a week, and sometimes two weeks, in order to avoid overshooting the birds. They hunted a course several different ways. Some courses had a start and a finish, and if they completed the hunt before they got to the end—that is, if they got the limit—they would sometimes start at the end the next time to try to even out the hunting. The goal was to take care of the birds, not to overshoot them. Most of the plantations had a ten-bird-per-gun limit with two guns, which is a twenty-bird limit per day. You could have four people shooting, but only two people shot at a time. The aim was to be conservative and not to overshoot the birds. It was a beautiful hunt.

The hunts also varied based on the personalities of the people involved and the quality of the dogs. To me, the best hunts were always the quieter hunts; I cannot stand a lot of screaming and hollering and cowboying, which people will sometimes do. Most people tried to hunt relatively quietly, though. The experience was not just in killing birds. The value in it was the total experience, and that meant the people, the birds, and the natural surroundings. Today, people are so obsessed with the numbers of quail they flush and shoot that they lose sight of the total experience. Mr. Stoddard always said that one bird per acre was a reasonable stocking on well-managed quail land. That way you would not have to sacrifice anything else, ecologically or aesthetically, to the production of quail. Since then, people have figured out that they can increase quail populations if they cut most of their timber and fill the land with planted feed. They plow up native ground cover, and the lack of pine timber depletes the fuel supply for prescribed fire. In time, they will not be able to burn regularly, and they will then lose their ecological and aesthetic diversity. If you are not interested in the total experience, you can hunt through cutover land, but that experience is not nearly as good as hunting through beautiful pinewoods with a healthy understory. I always thought of the hunt as a way of coming to understand the aesthetic of a healthy longleaf-wiregrass landscape. That is certainly how the hunts must have been when I was young, which was not only the heyday of these hunts but a time when these lands were in good shape.

Relations between the locals and the plantation owners were generally good back then. I know that my family, for instance, had absolutely no ill feelings toward our neighbors who owned quail preserves. There was very little animosity toward the preserves, though in a small community like Thomasville somebody always wanted to change something. We had a local man years ago who tried to get the county commission to raise taxes on the plantation land. Well, that did not stand a chance, because the plantations were the biggest industry that Thomasville had. A lot of merchants in Thomasville did very well, because plantation owners and their guests traded there, and the local people were not willing to alienate them.

In the early days, most of the preserve owners were nonresident owners. Even though some might spend more than six months out of the year in Thomasville, they still declared their residency up north, deliberately. They were not interested in participating in the political life of Thomasville, but they did spend a tremendous amount of money in Thomasville. In general, the owners contributed to the community, and they recognized an obligation to be sure that they were good citizens. I admired the preserve owners when I was growing up, mainly because they had such beautiful land and they took such good care of it.

I went into the army and then to college before I started my life's work. I joined the army a week after I graduated from high school in 1945. The war in Europe was already over, but they were anticipating an invasion of Japan. I went to Fort Leonard Wood in Missouri for my basic training. There was no question we were being prepared for an invasion: everything in our training was geared toward Japan. On my last day of basic training we were bivouacked in the woods, and they came out and announced the atomic bomb had been dropped. I was fortunate that I did not have to go through an invasion. I certainly admire all the guys who fought in that war. I stayed in the army for another year and a half, during which time I mostly processed returning veterans in separation camps.

When I got out of the army, I did not know a whole lot about college, just that I wanted to go and that I wanted to study forestry. Any profession that would put me in the woods sounded good to me. I had a friend in Thomasville named Coleman Carr. Cole and I were big buddies growing up, and he had a brother, Dick, in law school at the University of Georgia. Cole had just gotten out of the navy about the same time that I got out of the army, and we talked about going to school at Georgia. At that time, freshmen usually were sent to Savannah because they did not have enough space in Athens. I am sure they were culling out people who really did not want to be in school. Cole contacted Dick, and Dick told us to come on up and take the placement tests. If you passed all of them, you got fifty hours of credit, which was enough to enter the university as a sophomore. So Cole and I went up to Athens and took the tests, and we both passed them, so they accepted me as a sophomore. My first class at the University of Georgia was the winter quarter of 1947.

There was a tremendous number of returning veterans at the University of Georgia in the late 1940s, most of them on the GI Bill. One of my fraternity brothers had been a wing commander in the air force during World War II and was discharged as a colonel. A lot of the returning veterans, of course, had had terrible experiences in the war. We tend to forget that about World War II.

The University of Georgia's Forestry School was a new kind of experience for me. I had never been to college and had no idea what the curriculum would be. I took all the core university requirements and then got going with the required forestry courses. There was a course on forest fires, but it did not have anything to do with fire ecology or controlled burning. It was all propaganda about how destructive fire was. The other courses were pretty standard for forestry programs—mensuration, forest pathology, silviculture. I enjoyed taking ornithology under Dr. Eugene Odum, the leading ecologist of the day, who I found out later was a close friend of Mr. Stoddard's. They worked together on several projects. Later on, when I worked with Mr. Stoddard, we visited Dr. Odum in Athens several times, both at his house and at the university. He and Mr. Stoddard respected each other, because they were both smart, even though Dr. Odum was an academic and Mr. Stoddard was not. I took botany with Dr. Wilbur Duncan, who became one of the leading botanists in the Southeast, the author of several important guidebooks, and a strong conservationist. He was a young man when I took his class, but he was a tough instructor and a fine botanist. Dr. Merle Prunty was my geography professor, and I admired him as well. He gave a presentation at one of our Tall Timbers conferences later and did a lot of good work on the hunting preserves.

This was just before the pulp and paper companies started influencing the forestry school. They did not have much influence then, but it was just around the corner. Of course, the pulp industry was already in Georgia, mostly in Savannah and Brunswick. Down in the Red Hills, people who cut timber for the pulp market dealt with two mills: International Paper Company in Panama City, Florida, and St. Joe Paper Company in Port St. Joe, Florida. That was where most of the pulpwood from Thomasville went. But they did not exert much of an influence over the University of Georgia at that time. Brunswick Pulp and Paper and the company in Savannah, which was Union Bag in the early days and then Union Camp later on, soon exerted substantial influence at the university.

Pulp and paper companies eventually had significant power in most southern forestry schools. They had indirect influence by controlling the curricula at the forestry schools of the Southeast, including Auburn, Florida, and Georgia, and they eliminated conservation-oriented forestry for a time. They were only interested in production of pulp and paper for their businesses, and the college curricula reflected that. When I was working with Mr. Stoddard later on, we dealt with the pulpwood companies a good deal. It helped us in our thinning operations and our cleanup to be able to sell pulpwood. But we certainly did not grow trees just to supply the mills.

There were very few jobs available when I graduated from the forestry school. Some foresters were getting utilization jobs with sawmills, usually cruising or buying timber. The Georgia Forestry Commission also started expanding. They wanted a ranger in just about every county, so that furnished a lot of jobs for foresters. And then the paper companies were beginning to have an impact. Strangely enough, however, they did not hire many professional foresters. St. Joe, for example, had just one or two professional foresters for a long time. They owned a million acres of land and only had one or two graduate foresters. My friend Angus Gholson, an outstanding botanist, was one of their early foresters. He got a degree in forestry from the University of Florida and went to work for St. Joe Paper Company in the north Florida region, but he did not stay with them long.

Julie and I had gotten married in 1948, a couple of years before we graduated, and we were not sure what we were going to do. So we came back to Thomasville, and I went to work with my father doing several things. We grew some watermelons, cut a little timber, and planted some pines. Julie and I personally planted a lot of the pines on our place, with Julie driving the tractor and me sitting on the back planting. But it was just routine stuff. Mainly I was trying to work for our keep, because we lived there and ate there. I was looking for something better when I met Herbert Stoddard.

CHAPTER 2

Time Well Spent with Mr. Stoddard

When I was growing up during the 1920s and 1930s I was only vaguely aware of a man over in Grady County doing research on the bobwhite quail. My father, of course, was the land manager for his own property, and he was a quail hunter, but he was not a member of the Cooperative Quail Investigation, which had brought Mr. Stoddard to Thomasville. Nor was he acquainted with Mr. Stoddard personally. We were not part of the quail plantation group, and it was mostly the owners of the large plantations who funded and were members of the investigation. I do not think any local landowners were associated at that time, since most of them could not afford, financially or otherwise, to devote large pieces of land solely to the production of quail.

It was only when I went off to college that I learned more about Mr. Stoddard. I took the first course in wildlife management that was taught at the University of Georgia. Dr. Jim Jenkins, who was on loan from the Georgia Game and Fish Commission, taught the course; he became a full-time member of the faculty later on. Ed Komarek, who along with Mr. Stoddard would be a founding member of the Tall Timbers Research Station, was on the Board of the Georgia Game and Fish Commission back then, and I am sure he had something to do with releasing Dr. Jenkins to teach that class. Anyway, when Jenkins called the roll on the first day of class, we gave our names and where we were from, and I said I was from Thomasville, Georgia. Dr. Jenkins said, "Oh, wow, you must know Herbert Stoddard!" Well, I had never met Mr. Stoddard, but I did not say anything. I just smiled, so he always thought I knew him. I had it made after that to a certain extent, though I also had to be more prepared in class. He always called on me since I supposedly knew Herbert Stoddard, and I think he expected a bit more of me.

Herbert Stoddard was not only an influential figure in the early history of wildlife management in this country, but he and a small group of government and university scientists were largely responsible for the early teaching of wildlife management at universities all over the nation. In the late 1920s, when Mr. Stoddard was working for the U.S. Bureau of the Biological Survey, he also worked with Aldo Leopold to set up a series of graduate fellowships at four universities. The funding for this initiative came from the Sporting Arms and Ammunition Manufacturers Institute, or SAAMI, an industry group that was led by John Olin, who much later came to own Nilo Plantation in the Albany area. We worked with Mr. Olin on Nilo until he died. Stoddard and Leopold scouted out promising young students with an interest in wildlife management—it was known as game management back then—and placed them in university jobs to teach and conduct field research based on the model Mr. Stoddard developed in the Red Hills. Paul Errington ended up being the most influential of these wildlife ecologists.

His early research was on bobwhite quail in Wisconsin, and he provided some of the first scientific evidence that predators were natural and important parts of ecosystems, rather than something for people to exterminate. He went on to teach wildlife ecology at Iowa State University and become an important spokesperson for nature. These fellowships set up by Stoddard and Leopold gave wildlife management some legitimacy in the university setting. When I took Jim Jenkins's first wildlife class at the University of Georgia, then, I was the beneficiary of the pioneering work of Stoddard and Leopold.

I finally met Mr. Stoddard in a roundabout way. While I was in school, I thought I had a job lined up with the Georgia Game and Fish Commission upon graduation. I wanted to get into the wildlife field if I could, even though I took a forestry degree. But Georgia's government was in the middle of one of their typical messes. The 1946 governor's race pretty much had it all: three candidates claiming they had won, fabricated votes, the physical seizure of the governor's office and residence, and the death of Gene Talmadge. It is a far more convoluted story than I know how to tell. The point for me is that the fallout from that race caused my job to vanish before I even graduated.[1] Instead, I came home with Julie, my beautiful bride, and moved in with Mother and Daddy. I helped my father on the farm while I was looking for a job.

I did not have much success in my job search for a while, but one of my classmates from Albany, Robert Tift, landed a summer apprenticeship with Mr. Stoddard. Robert was the half-brother of Richard Tift, who worked closely with Mr. Stoddard in land management and plantation development in the Albany area. Richard had made arrangements for Robert to apprentice under Mr. Stoddard that summer after we graduated. Robert did not have a place to stay, so he moved in with us at my mother and father's house, in the spare bedroom. Robert went out with Mr. Stoddard every day, and at some point I asked if it would be possible for me to go along. I had wanted to meet Mr. Stoddard, because I had heard so much about him in college. Robert made the arrangements, and so I started going out with them every once in a while. Then I started going out whenever I could, which was maybe once or twice a week, whenever I could get away.

I always had an interest in nature—in the woods—and I tried to learn more about the natural world every chance I got. I could have been a better naturalist if I had had more sense, but I had a few bird books, and I could identify most of the region's birds. I also knew animals, how to hunt and fish, and how to read what was happening in the woods. And I guess you could say I was inquisitive, and that came out when I was in the woods with Mr. Stoddard and Robert. Mr. Stoddard, who had also grown up in the woods in Florida, apparently took note of that. He especially appreciated that I had an interest in birds, which were a passion of his.

Mr. Stoddard had only agreed to keep Robert on for the summer, and Robert got a job with a paper company over in the Savannah area when the summer ended. So I knew that Mr. Stoddard was going to need some help. I went out with Mr. Stoddard a few more times, and finally he said, "Have you gotten a job yet?" I said, "No sir," and he said, "Would you consider going to work for me?" I tried to act like I would give it some thought, but I was thrilled with the offer and I jumped on it then and there. He hired me by the hour, and only when he wanted me to work, which was fine by me. I would have paid him to work with him if I could have afforded it. That was in 1950, and it was the beginning of a wonderful relationship.

Not long after Mr. Stoddard hired me, Julie and I needed a new place to stay. We had been living with my mother and father, and we were ready to get out, as one might imagine. Mr. Stoddard had a little tenant house at Sherwood Plantation that stood empty. He said we could move in there if we wanted to, and of course we did. We had to run water and electricity in there and do a lot of work on the house, but we eventually moved in. We ended up living there for twenty-one years, until Mr. Stoddard died.

Sherwood was an interesting place. Lewis Thompson, one of the leading figures behind the Quail Investigation, had given it to

FIGURE 2. Herbert Stoddard, right, hired Leon Neel, left, in 1950, and together they fashioned a unique method of forestry and land management now known as the Stoddard-Neel Approach.

Mr. Stoddard in 1930. Mr. Stoddard was already looking for some land in the Thomasville area after the study was over, and when Thompson heard that, he thought the gift of Sherwood might entice Mr. Stoddard back to the area permanently. He was right as it turned out. "Colonel" Thompson, as he was known (it was an honorary title given him by the governor of New Jersey), owned a large plantation called Sunny Hill on the Georgia-Florida line, and he had used Sherwood as what was called a "satellite" plantation. It was primarily managed for quail at the time. But once he took control of the land, Mr. Stoddard was not managing Sherwood as a quail preserve, because quail production was not his primary goal. He was growing quality timber, and he was maintaining diversity for the maximum number of species on the land. He also had an annual cash crop when we moved out there. He had about a hundred acres of tung trees that produced nuts. The nuts were gathered and sold, and the oil was used in the manufacture of paints, varnish, and munitions. In fact, he had been using the tenant house to store tung nuts before we fixed it up and moved in. He also ran cattle on the place. He had a thousand acres in four land lots, which formed a square. It was fenced on the perimeter all the way around, and he had an arrangement with a local cattleman in Cairo, Georgia. At the right time of the year they would put a hundred steers on the place, and they would free range until they fattened up on the woods pasture and the fields. The tung fields had crimson clover, which fixed nitrogen and provided graze for the cattle. Then he would round up the cattle and sell them. So he had several cash crops on the land.

Yet despite all of these activities, Mr. Stoddard maintained Sherwood's biological diversity. It was rolling Red Hills land, not too far from the Ochlocknee River. It had two or three good creeks on it, as well as three or four ponds that Mr. Stoddard built. It was rich game land for turkey and deer, with some quail, of course, because he burned his property. It was a wonderful place and still is. His son and grandchildren have taken good care of it.

When I went to work for Mr. Stoddard, he was doing very little laboratory work indoors, except for preparing bird skins. He was an avid taxidermist who had studied with some of the best museum people of the era, and had worked at both the Milwaukee Public Museum and the Field Museum in Chicago. Back then, taxidermy was an important natural history technique, and Mr. Stoddard was tremendously skilled at it. When he moved down to the Thomasville area, he kept at it. He had a room at Sherwood that was dedicated to preparing bird specimens.

There had been little professional ornithology practiced in this area, and so Mr. Stoddard collected birds to expand ornithological knowledge about the Red Hills. In the time I knew him, he collected quite a few birds, and he sent most of his skins to the National Museum of Natural History in Washington, D.C. I watched him skin birds for many hours while we talked over things in his skinning room. Visiting ornithologists would always

FIGURE 3. "The Hall" at Sherwood Plantation was the family home of Herbert Stoddard from 1924 until his death in 1970. Landowner Lewis Thompson gave Stoddard the house and the surrounding thousand acres in 1930 to entice him to stay in the Red Hills, and it became a major destination for the nation's most distinguished naturalists.

comment that Herbert Stoddard made the most beautiful bird skins that they had ever seen, and those skins became important resources for ornithological study. He did not do much other lab work. The field was really his laboratory. He was always experimenting in the field.

When I began working for Mr. Stoddard, I focused mostly on forestry. That is what he groomed me for. He taught me how to mark timber and the whole process of forest management on a quail preserve. Mr. Stoddard had a fierce work ethic, and working with him was no picnic. Ed Komarek once wrote, about Mr. Stoddard, that "time and hours were of no consequence, and anyone that worked with him had to realize this or lose his respect."[2] Ed was right about that. Time meant absolutely nothing to him when he was on a project. It did not matter what time of the day or night it was, or what the weather was like, or anything else. If he had something in mind to do, we were going to do it.

When it came to marking timber, we had a daily routine. I lived just across the road from his house about a quarter of a mile away, so that made things convenient. He would come to pick me up when he got ready to go to the woods, usually about 8:00 a.m. He wrote his correspondence in the early morning and caught up on other paperwork before heading out into the field. He was also waiting for the light to get right. You do not mark timber in the dusk of late afternoon or the twilight of early morning, and you do not mark timber in the rain either, because the rain changes the appearance of the foliage to the extent that you cannot see the needle color. Color is very important. The needles begin to turn yellow on a weak tree or an old tree showing weakness. You can also make marking judgments based on the number of needles in a cluster. In the rain, or at dusk and dawn, you cannot see these things to gauge the health of the tree. All of these things are extremely important in choosing which trees to remove.

Mr. Stoddard used to talk about his early experiences when we were out there in the woods. His childhood years in central Florida remained influential throughout his adult life. For example, when we went on several ivory-billed woodpecker exploration trips, he would often refer back to when he was a boy in Seminole County, Florida, before 1900. He saw ivorybills down there, right around Lake Mills where he lived, and he could recognize good habitat for them. He always relied on his experience in the woods. Because he never finished high school, let alone went to college, his childhood in the woods was his formal education.

His early experience as a boy in the Florida woods, before his family moved back to the upper Midwest, was particularly important in regard to his understanding of fire and its ecological and cultural place in the longleaf region. But his first experience with fire management, like mine, was not particularly auspicious. When Mr. Stoddard was just five, he decided, after watching the local cattlemen burn the woods, that he wanted to try it himself. So he set a grass fire near his house on what turned out to be a dry, windy day, and that thing apparently took off. The neighbors acted quickly and barely saved the Stoddards' house, but that fire burned for three days through the adjacent woods.

The cattle people he was associated with in Florida routinely burned the woods, and he learned from that. Cattlemen set fires deliberately, not only to improve grazing but to concentrate cattle in particular places. This was one of the ways he came to see fire as an important management tool. There is no question that his experience in Florida affected his land management decisions as an adult. In his autobiography, *Memoirs of a Naturalist* (1969), he tells about being a sort of gofer for a cattleman named Gaston Jacobs, who controlled a lot of livestock over a big acreage in Seminole County, Florida. He went on cattle drives, or roundups, with Jacobs and his family.[3] Mr. Stoddard, of course, had a good mind, and he absorbed everything he saw. The burning that these cattlemen did was not quite like what we do now, or what Mr. Stoddard did in the 1920s and 1930s. The range was so big and open at the turn of the last century, though, that they could light a fire and let it go until it went out on its own. They did not have to be as careful back then as we do now. Moreover, I doubt they really understood the ecological benefits of burning the forest in the ways that we have come to understand them. But they did know fire was necessary to keep the grassy understory in good shape. Fire kept the forest open, and it encouraged good, fresh grass to grow. So that was, in many ways, the launching point for Mr. Stoddard's long-term investigations of the ecological and historical role of fire on the southern coastal plain.

Another important aspect of Mr. Stoddard's learning process was that he always reinforced his early beliefs through subsequent experiences. When he came to Georgia he was mainly working on the Quail Investigation, but he also found the time to do a lot of outside work for the Biological Survey as well. He started going to the Gulf Coast, where he was involved in the development of the St. Marks Refuge. The St. Marks National Wildlife Refuge was established in 1931, and it is an interesting mix of coastal marsh and pine flatwoods. Mr. Stoddard wrote several early reports on the area for the Biological Survey in the 1920s, describing its flora and fauna as well as assessing the political feasibility of creating a wildlife refuge in the area. There were free-ranging cattle south of Tallahassee in the flatwoods back then, and so he felt right at home in that environment. It reminded him of his Florida childhood. The Pinhook River was on the refuge, and he went down there with Sid Stringer, who was the manager of Colonel Thompson's property where Mr. Stoddard lived, to go hunting and fishing. But Mr. Stoddard was also an ornithologist and was always collecting birds. The area in and around St. Marks was burned regularly for cattle, which made it ideal for all sorts of wildlife. He told me many stories about how they used to burn so regularly down there that, as a result, there was wonderful hunting along the coast. Jack snipe and geese were favorites among many game species. More importantly, down along the Gulf Coast he saw landscapes that were managed similarly to those of his childhood, and with great ecological results. So he brought that experience back to reinforce what he was learning about fire management in the Red Hills.

There was another fellow down in central Florida, an elderly neighbor he remembered only as Mr. Barber, who got him interested in natural history when he was a child. Mr. Barber, who was a retired government surveyor, often took young Herbert fishing, and in the process he taught him all about the local wildlife. He was instrumental in Mr. Stoddard's development as a naturalist. Years later, Mr. Stoddard dedicated his *Memoirs* to "Mister Barber." Mr. Stoddard had a mind that would absorb anything about the natural world, because that was the only world he knew, and Mr. Barber fed him a steady diet of new information.

When he moved back to the Midwest as a young man, he came to know many influential naturalists up there as well. After his family relocated, he taught himself taxidermy, and he had a couple of drugstore displays in Rockford, Illinois. He dropped out of school soon afterward, and he went to work on his uncle's farm in Prairie du Sac, Wisconsin. He had to work for his living, and his living was nothing more than food and shelter. That was where he got his early start in professional taxidermy with Ed Ochsner, who became a lifelong friend and mentor. During the winters, when the demands of the farm slackened, Ochsner taught Mr. Stoddard how to prepare animal skins. They also lived near Baraboo, Wisconsin, where the Ringling Brothers Circus had its winter headquarters, and when Mr. Stoddard was a young man, he and Ochsner once paid a visit. While there, they heard about the death of a hippo, a big bull, and Ochsner apparently convinced Arthur Ringling to donate the animal to the Milwaukee Public Museum. The museum sent its head taxidermist, George Shrosbree, over to prepare the hippo, and Mr. Stoddard stayed on to assist Shrosbree. The whole process took a week, and Shrosbree must have been impressed with Mr. Stoddard's work, because the Milwaukee Public Museum soon offered Mr. Stoddard a job.

It was probably up in Wisconsin that Mr. Stoddard started to develop a true sense of what his friend Aldo Leopold called a land ethic. In particular, Mr. Stoddard did not like waste, an aversion that came from growing up in hard times. He could not stand how wasteful modern society was. People would probably laugh today, but I remember we were in the woods one time, and we came upon where somebody had dumped some trash. Now he detested that, but he was not merely angry that they had littered. They had also dumped several perfectly good quart glass jars in the trash. Mr. Stoddard said, "Look at that waste. When I was young my mother could have used those jars for all sorts of things, and would have felt lucky to have them." So it really got to him that objects like that, useful objects that had value in austere times, were becoming disposable. And his feelings about waste relate directly to what he did with land. He could not understand why someone would clear-cut a stand of timber just for the money. That was not only a waste of ecological resources, but it also traded off long-term income for a short-term gain. It was like throwing away so much of what was useful and valuable out there in the woods.

While Mr. Stoddard was working for the museums, he made a name for himself. The more his colleagues and peers got to know him, the more they admired and accepted him. These connections were important, because nobody could have accomplished what Mr. Stoddard did—building a tremendously successful professional career as a scientist, forester, and land manager without so much as a high school diploma—without the help of colleagues and personal connections. You simply could not pull that off today.

Julie and I, in the early days, met several of Mr. Stoddard's closest friends from his museum days. One of them was Owen Gromme, who replaced Mr. Stoddard as associate taxidermist at the Milwaukee Public Museum and went on to become a well-known wildlife artist. Mr. Stoddard actually helped Owen get started on his art career. Julie and I got to know him, his wife, Ann, and his family quite well. He had a son and a daughter. The daughter was the youngest and she was an excellent fly-fisher. Mr. Stoddard had a pond on the place and I would take her fly-fishing. We would set out at first daylight, and she would catch some of those big bluegills on a top fly. Clarence Jung, an ornithology companion from Milwaukee, also used to come down, and others that I have forgotten. Folks like this maintained their friendships with Mr. Stoddard throughout their entire lives.

W. L. "Waldo" McAtee, who was Mr. Stoddard's boss at the Biological Survey during the 1920s, used to visit a lot as well. He was a wonderful man and an outstanding ornithologist. He would come down and spend a few days with Mr. Stoddard to talk over anything of mutual interest in the conservation field, and to collect birds. Ornithologists rarely collect birds today, but in those days—this was in the 1950s—if you wanted to claim a record for a rare bird, you had to have the feathers. So it was a common thing for scientists to shoot birds; you had to have a collector's permit, of course, but most working ornithologists collected specimens. Stoddard and McAtee would go out collecting whenever McAtee visited.

One time I remember going over to the office, and he and McAtee were in there examining a few nighthawks. It was after dark, and they had just collected them over on Ed Komarek's pasture. They

FIGURE 4. As the head of the Food Habits Research division of the Bureau of the Biological Survey in the 1920s and 1930s, Waldo Lee McAtee was an important early influence on the development of wildlife biology and wildlife management. He was also a close friend of and advocate for Herbert Stoddard.

were looking for the Antillean nighthawk. McAtee had the feeling this particular species might be in the Southeast on occasion, so they collected a series of five or six. The only way you could tell the difference between the Antillean and the common nighthawk was by comparing the skins. Well, I went over there and they had taken the crops out of all these nighthawks, and they were examining what was in the crops. It turns out that they were all full of the same little tiny beetle. There were dozens and dozens of beetles; that was all they were eating. I walked over there and said, "What have you got here," and Mr. Stoddard said, "Well, this is such and such a beetle." They were a species of dung beetle. And being even more uninformed than I am now, I said, "Boy, you can sure tell what they like to eat." McAtee raised his head real quick and said, "Wait just a minute. That is the wrong way to think about it. You better be saying you can really tell what was *available* for them to eat." That was one of my first wildlife management exams, and I flunked! But he was right. They were not just going around picking those beetles out of a bunch of other beetles they could have eaten. They were eating what was available. And I have checked the validity of that point several times since then, not with nighthawks, but with other species. Generally I find that something has been eating something else not because they necessarily preferred it but because it was available.

Dr. McAtee had a lot to do with Mr. Stoddard getting the job with the Biological Survey that resulted in his landmark quail study, *The Bobwhite Quail: Its Habits, Preservation, and Increase* (1931).[4] As he worked in the museums, Mr. Stoddard was also making a name for himself as an ornithologist. McAtee, along with S. Prentiss Baldwin—an important ornithologist who had perfected bird-banding methods in the 1910s—had watched Mr. Stoddard coming up through the ornithological societies and thought he would make a pretty good lead investigator. So McAtee convinced E. W. Nelson, the chief of the Biological Survey, to hire Mr. Stoddard for the Quail Investigation. Mr. Stoddard accepted the offer, and he came down to Thomasville in early 1924. It was a sort of homecoming for him.

When Mr. Stoddard arrived in the Red Hills most of the prominent quail plantations were already in existence, and the oldest ones had been around for thirty or forty years. The land, even the land that was not in plantations, was mostly farm- and timberland. Timber had very little value in those days, so the timberlands that were there generally had not been abused. Mr. Stoddard always said that Thomas County was heavily forested in those early days. The fields were scattered throughout the timberland, except across the Florida line, which had more agriculture. Of course, the agriculture on the Georgia side did not look like it does today. The fields were much smaller, and they left hedgerows and other wildlife cover all over the place. By that time there were already some fields that had recently been abandoned, so plant succession was coming in with a few scattered pines, which made excellent quail habitat. As cotton declined in the region, the amount of such early successional habitat continued to grow. There were many more old-growth stands of timber like the Big Woods at Greenwood Plantation back then, though the overall volume of timber was probably less than it was half a century later, after we had practiced the Stoddard-Neel system on many of those properties. But it was good land with good stands of timber nonetheless.

A group of plantation owners had called in the Biological Survey after World War I because of an apparent decline in the quail population. There was still a lot of good habitat, but some of the plantations had stopped burning their land. The blame for that has been put on the anti-fire movement spearheaded by the American Forestry Association and the U.S. Forest Service, which began after World War I and lasted on up until the 1930s. Until then, people in the region had been burning everything that would burn, more or less, but when professional foresters got wind of such practices, they were aghast. They were convinced that burning was destructive of forests in general and a pox on the land. When the Forest Service and other forestry organizations implored people to stop burning, some did stop, and that degraded the area's quail habitat. It is important to realize that good quail numbers up to that time had been an unintentional byproduct of patch farming and traditional woodland burning, not a result of conscious wildlife management.

In retrospect, it is difficult to pinpoint one single factor that led to the quail decline, though fire suppression must have been an important one. But there is one thing I do know: you cannot maintain a stable *wild* quail population on any land. The population is going to fluctuate no matter how much you try to control it. High predator populations, drought, and a host of other factors can affect quail populations. After twenty years or more of good quail hunting, the population might have decreased for one reason or another. The decline of burning was certainly part of it. I do not want to deemphasize that, especially in the woodlands. If they quit burning their woodlands, a pretty thick rough would have grown up in many places, and then they probably lost quail

FIGURE 5. Herbert Stoddard developed and improved many techniques for conducting wildlife research in the field. Here he is banding a bobwhite quail in 1924, the first year of the Cooperative Quail Investigation.

in there. You certainly cannot maintain many quail in woodland habitat without burning. Of course, the Quail Investigation is best known for what Mr. Stoddard had to say about fire, and how most professional foresters thought he was nuts to advocate its use, but there were almost certainly other factors at play, which is why Mr. Stoddard took such a broad approach.

Regardless of what the problem was, I think we can all agree that it is a good thing they recognized that there was one. Mr. Stoddard's quail study, the eventual product of that concern, was the first systematic life history study of an upland game bird to determine its habitat preferences—what these birds liked and what they did not like. The importance of this study cannot be overstated. He not only conducted a life history of the bird itself when he studied the quail, but he did the life history in relationship to its environment, particularly dealing with population dynamics. And he showed how landowners could do small but important things to improve habitat by, for instance, increasing food supplies and providing adequate cover for the birds.

Let me give you an example. The Apalachicola National Forest is a wonderful piece of land, but they have a fairly small quail population down there. It is down southwest of Tallahassee, and it was established in the 1930s on land formerly used for timber and turpentine production. It is Florida's largest national forest at over five hundred thousand acres, and has several stands that contain old-growth longleaf. They have quail in the forest, but not in shootable numbers. With some land management, following some of Mr. Stoddard's recommendations, you could have more quail in there if that is what the managers wanted, but that is not what they want. They could do it without really sacrificing other values, but it would take some work and money.

Through the quail study, Mr. Stoddard learned how to systematically manage the environment toward a specific end—which initially, for him, was a healthy quail population—without destroying the aesthetics of an ecologically healthy forest. We played with that idea continually over the years, all the while adding new economic and ecological factors and concerns. Mr. Stoddard began to broaden his management to where, in our Stoddard-Neel system, we manage for as much biological diversity as we can while also providing some economic dividends. As the quality longleaf land base shrank, we realized that, as we were managing and manipulating land for specific ends, we could also broaden the effort and learn how to manage for the benefit of the majority of species, even while emphasizing one or two like timber and quail. That goal is at the heart of the Stoddard-Neel Approach, and many of those insights first began to emerge in the Quail Investigation.

It is interesting to read Mr. Stoddard's quail book today, considering all that we have learned about the longleaf-wiregrass forest and all that has happened to it since then. He was obviously in tune with the forest, and he knew how it worked. But he did not have much to say about wiregrass, which, for ecosystem managers today, is one of the key indicators of an undisturbed and healthy forest. I think there were several reasons for that, starting with his focus on quail. He remarks in his study that you had to burn the heavy stands of wiregrass because quail would not go into them if you did not. He even advised landowners and managers to plow strips through it, which he would never do today. When he was working that land in the early days—even when I started working for him in the 1950s—there was a larger acreage of wiregrass and longleaf than there is today. It was easy not to worry about something when there was plenty of it. There was a lot more undisturbed land back then, with a lot more wiregrass, and that could be a challenge for quail management because quail do well on disturbed land if all other factors are in place. It was only when they started clearing land again on a larger scale, and making larger fields, that quail began to decline in numbers. And now many of the quail management people are harrowing up wiregrass. They are doing a lot of damage now that they should not be doing, all in the name of producing more quail.

Another interesting thing about *The Bobwhite Quail*, in comparison with how the Stoddard-Neel system has evolved, was Mr. Stoddard's treatment of fire. The type of fire he writes about is a bit different than what we practice today. This was partly a result of the compromises the Forest Service forced him to make, but it also came from the singular focus on quail. In the book, he suggests using low-level winter fires only, a burning regime that was 100 percent in favor of the bobwhite quail. We try to mix up our season of fire a little more now, especially when we are trying to capture longleaf regeneration or control hardwood encroachment. Mr. Stoddard was still working some of these things out when that book went to press. If you read all of his publications closely, you will notice that he always included a statement along the lines of "our present state of knowledge tells us" such and such. He knew that people would continue to learn more—the forest is never in a fixed state, and neither was his science. He may have understood the complexities of the longleaf forest at some level, but his land management had not yet developed to include everything.

The Bobwhite Quail was a tremendous achievement. But we have to remember that he continued working in and learning about longleaf woodlands for another forty years after that book appeared. The book represents a snapshot of his learning process at a particular point in its maturation. When he first came to the Red Hills in 1924, he had not been working in the South. In fact, he had not been in the South since 1900. He was working for the Biological Survey and for the quail plantations to develop quail, so that was his focus at first. He evolved just like we all do. In fact, this constant learning process is one feature of the Stoddard-Neel Approach. Since we do not have rigid formulas, we are free to adapt. The point I am making is not that Mr. Stoddard missed things about the longleaf-grassland system; it is that learning over time is built into our approach.

Of course, Mr. Stoddard's advocacy for and use of fire was a revolutionary aspect of his quail study, but I would go further to say his entire system of management was revolutionary. It was not just fire. Take his view of predators, for instance. He was one of the first to learn how to minimize certain predators through managing habitat rather than trying to eradicate them. Paul Errington, in fact, followed Mr. Stoddard's lead in his more thorough study of predators in the early 1930s. The predators that Mr. Stoddard feared most were the Cooper's and sharp-shinned hawks. They are the greatest avian predators on quail. In fact, it was the quail study that really identified these two hawks—together they are often referred to as blue darters—as the main predatory threat to adult quail. It was one of the most important findings of the study, and it let a lot of other alleged predators off the hook. Cooper's hawks can be devastating, especially in the spring migration. We generally burned in the spring after the quail season, which happened to coincide with the northward migration of the Cooper's hawk. They can kill many quail during this period. When Cooper's hawks come through this region, they do not just fly over. This area has plenty of their favorite prey, so some stop off to work the land.

This reminds me of an interesting anecdote: I had the privilege of shooting at Ichauway for a few years. On one hunt, Jimmy Atkinson, the outstanding land manager at Ichauway who is also my close friend, was in charge of the hunt. Jimmy was on horseback, and I was riding the wagon. We rode along, and Jimmy pointed out a puff of quail feathers on the ground where a Cooper's hawk had eaten a quail, or at least had picked a quail. He said, "That's the second or third one I've seen this morning. We've got a lot of Cooper's hawks in here now."

Soon thereafter, we had a point up ahead on old-field land with a strip of old, planted slash pine. It had grown up to blackberry briars, and they had cut trails through them with the mowers. Jimmy was standing by me, and we were watching the guns go up to the point. From underneath the briars on one side, an adult Cooper's hawk walked out into the opening and took off. Now, Cooper's hawks are not like a big old soaring hawk. You can see them, but you have to be looking to see them. A little while later, I was in the wagon and Jimmy was off on another point. I just happened to turn around, and I was looking back down the trail from where we had just come. About a hundred yards down there, another Cooper's hawk walked out of the briars and flew off. Twice that day I saw them walking out of those blackberry briars. They were in there trying to flush some quail out. They had seen quail go in, and they went in after them. I think we counted eight or nine places on that particular hunt where a Cooper's hawk had picked a quail. The point is that cover is vital to quail survival, but a Cooper's hawk on certain occasions will go in there and flush them out no matter how good the cover. When there are Cooper's hawks around, the quail often have no place to hide.

Still, even with the blue darters, Mr. Stoddard believed in controlling predators through habitat management, not direct eradication. That is why he developed those little circles of cover in open pinewoods. He liked to plow a firebreak around good patches of cover like plum thickets just to leave quail somewhere to hide. I have seen Cooper's hawks miss their target just as often as they hit it. Mr. Stoddard and I were marking timber on Foshalee Plantation one time years ago, and we flushed a covey of quail. They got up and took off and, out of the blue, a Cooper's hawk came in behind them. One quail dove into a little thick quail covert, not as big as a small living room. He went in there for his life and that hawk was right behind him. But the quail came out the other side, and the hawk never did come out while we watched. He was in there trying to hunt him out. That is how the quail escaped, not by going in there and staying. He threw the hawk off just enough. Now that does not always happen; the hawk will sometimes see the quail and get it. But Mr. Stoddard believed in providing enough cover habitat to give the quail a fighting chance.

The difficulty with controlling animal predators is that their populations fluctuate. When there are a lot of them, they can be bad; when there are not so many they are obviously not a problem. But if you eliminate one predator, you are probably favoring another one, which can have larger ecological ramifications. Foxes, for example, can play havoc with young quail, and at one time years ago, they were the main predator on Nilo Plantation in the Albany region. The owner, John Olin, ran a very tight ship on quail management. Mr. Stoddard worked with him on overall management, and I worked with him on the timber primarily. Mr. Olin had a crew that gathered information on quail throughout the year. They monitored nesting activity, recording the clutch size, hatching data and predator activity, and during the hunting season they recorded the crop content, sex, and size of each quail bagged. At one point, they determined that foxes were the main mammalian predator, so Mr. Olin instigated a severe trapping program, and he knocked the fox population down. There was little true woodland on Nilo; it was more a field environment. So it was easy to do that up there because of the cover. Well, the next year the cotton rat population increased, and he lost just as many quail. It turns out that the foxes were controlling the cotton rats, and once you reduced their population, then another problem popped up.

Mr. Stoddard had already been through situations like that during the Quail Investigation, and he knew better than to guess about something as complicated as predator-prey relations. That was something that required intensive study. So he took a broader view of predator control. Even with Cooper's hawks, while he was wary of them, he would never argue for outright elimination. In the first place, he felt like nobody could do it. It would be a waste of time and money. Instead, he was constantly considering how you could reduce a predator population by better land management.

Mammalian predator control was also important, and it has changed a great deal over the years. When Julie and I moved out to Mr. Stoddard's property in 1950, you could step outside almost every night and hear a hound running somewhere. There were a lot of people still living on the land, and most of them had a hound and went hunting many nights during the cool months. There was still a pretty strong fur market back then, and a lot of people hunted small mammals for the pot as well. Mr. Stoddard liked that form of predator control because it was useful, not wasteful. It was not done out of fear or malice, but instead was another example of how human activity on the land could fit in with ecological processes. The locals kept mammalian predator populations in check, but they were not out to eradicate anything. With fewer people on the land, that form of predator control is not really available now.

This discussion of predators reminds me of something Mr. Stoddard always used to say: "Nature plays tricks on a man." What he meant is that human intelligence is not capable of understanding everything that goes on in nature. Often, he would discover an ecological process that he had not quite understood before. Of course, when he discovered something it meant that he said, "Wait a minute, maybe we're not doing this right; maybe we ought to do it this way." It was not black and white, wrong and right, but he knew something was not working right, and at some point he gained enough insight to say, "Well, maybe if I do this, it will be better." So he was constantly willing to change and, hopefully, improve his management practices. At the same time, he never did get mad about it. He would get a chuckle out of it. That was one of the things he meant, too. Nature was playing a trick on him, and with something as complex as predator-prey relations, nature could play an awful lot of tricks. Those tricks can turn ugly if you are arrogant enough to assume that nature is a simple thing. Every species plays its own important role in nature, and if you try to eliminate one, you are likely to create as many problems as you solve.

After the original Quail Investigation was over, Mr. Stoddard moved to Washington, D.C., to work up drafts of the book manuscript. He was up there for a year or two, and he still planned on working for the Biological Survey when he was done, but he was not a very happy man living in the city. He would have stayed there the rest of his life if he believed that to be his responsibility. But that man belonged in the woods, and I think he knew it. He had already decided that he might like to come back down to the Red Hills to start a consulting business. He asked Sid Stringer, who lived in the house at Sherwood and worked for Colonel Thompson, to look around to see if anything was available in the area. Well, next thing he knew there was a letter from Colonel Thompson offering to give him Sherwood. I doubt it took Mr. Stoddard very long to accept that offer. Shortly after the plantation owners heard about Mr. Stoddard's impending return, they decided to continue as an organization, this time separate from the Biological Survey. They called their new private group the Cooperative Quail Study Association (CQSA), as opposed to the Cooperative Quail Investigation. All of the members, who were again mostly quail plantation owners, paid dues to fund Mr. Stoddard's continued research and to retain him as a land management consultant. The CQSA was created before the government offered any advice on wildlife management, so it was an important resource for landowners. While the CQSA was centered in the Red Hills, its membership included landowners from the Carolinas to Mississippi, and so Mr. Stoddard got to visit and work on quail lands throughout the southern coastal plain. That association also laid the groundwork for the formation of the Tall Timbers Research Station, the subject of the next chapter.

As long as Herbert Stoddard was a federal employee, he had to temper his rhetoric about fire. Indeed, he had to repeatedly revise his fire chapter in *The Bobwhite Quail* to satisfy the concerns of U.S. Forest Service officials that his study not appear to be condoning widespread burning. But after Mr. Stoddard left the Biological Survey to become a wildlife consultant in the private sector, he could speak his mind on federal fire management policy. He used this new position and the freedom it afforded to fight even more vehemently for controlled burning. He had already gotten the quail plantations to start burning again, and that was during the heyday of the fire fight in the Southeast when the American Forestry Association had sent their "Dixie Crusaders" into the South. They traveled around the region from 1924 to 1928 to suppress woods burning, and they were pretty successful at spreading their message. It did not take long for them to hear about what Mr. Stoddard was doing in the Red Hills. And so, in 1928, some of them invited him to give a presentation at the regional meeting of the Society of American Foresters. It was in Jacksonville, Florida, and he gave a paper on woodland game management and talked about the benefits of burning the woods. It was almost as if they were setting him up. They were in no mood to hear that fire could be a good thing, so they pretty much booed him off the stage. That was one of his favorite stories because it illustrated how vehemently foresters opposed the use of fire in those days. While it took some time to achieve, Mr. Stoddard was one of several foresters and land managers whose advocacy of fire management finally produced a revolution in ecological management across the country.

Ed Komarek used to tell a funny story about Mr. Stoddard. Whether it is true or not, I think it illustrates the unique role that Mr. Stoddard played in mediating the relationship between local land-use traditions and professional land managers. It also illustrates how he liked to have a little fun at the expense of misinformed officials. There was a lot of sniping going on by the Forest Service toward Mr. Stoddard, of course, and they sensed they were going to lose the battle over fire, certainly on the game preserves. I think they resented that, so one of them came down from Washington sometime in the 1930s. He was not the Chief of the Forest Service, but he was a pretty high level official. Mr. Stoddard was riding him around and being very courteous, as he was to everybody who came by, even though he knew this official was just looking for an opportunity to challenge him on fire management. It was in the spring of the year, and they went through one of the plantations. The tenant system was still in effect, and one of the cabins was occupied by a very nice black lady named Miss Sally, as I was told. She had her own small house, and, like many folks did back then, she kept the yard swept around it. Then she had a flower garden all the way around, a couple of fig trees, and a vegetable garden. That was Miss Sally's home.

Mr. Stoddard and Miss Sally had become friends over the years. Every time he was close he would go by, and if she was out in the yard he would stop and talk to her. On the day the Forest Service official visited, they drove by, and Miss Sally was out in her yard. Not only that, but she had just burned around her house, as the plantations generally let the people control a certain parcel of land around their houses. Mr. Stoddard stopped, they got out, Mr. Stoddard introduced this man to Miss Sally, and they talked for a while. Then the Forest Service official said, "Somebody's been burning around your house here," and she said "Yes, sir." He asked, "Did the plantation do that?" And she said, "No sir, I did it. They let me do that, so I burned a couple of days ago." He said, "Well, look at it. You burned up everything out there. Just look at it. That land that you burned, it's just as black as you are." Miss Sally paused for a minute and then politely replied, "Yes sir, and in about three weeks it'll be just as green as you are!" Ed loved telling that story because it not only showed how misinformed many people in the

Forest Service were, but also how knowledgeable local people were about the usefulness of fire.

In the early 1930s, as the Cooperative Quail Study Association expanded and Mr. Stoddard's responsibilities increased, it became clear to him that assistance would be critical to his success. After a few years of consulting all over the Southeast, he was about to work himself to death. He was busy conducting experiments on Sherwood, writing up publications, and traveling all over the region to examine land. So he was lucky when, in 1933, the brothers Ed and Roy Komarek stopped by to visit Sherwood. Ed was a student at the University of Chicago studying under the famous ecologist W. C. Allee, and Roy was about to enroll there. They had been collecting mammals for the Chicago Academy of Sciences in the Smoky Mountains, where they had spent about a year. They lived in a little shack up there during the height of the Depression. They were on their way to do some work in the Everglades when they stopped at Sherwood.

Ed, being an energetic and progressive young naturalist—and a brilliant one as well—had read and heard about Stoddard. Ed had never met him or come into contact with him, but he knew of his history at the Field Museum in Chicago. They only stayed a couple of days and then continued on down to the Everglades. But Mr. Stoddard must have seen enough of Ed to realize that he would make a good assistant, and so when they came back through a few months later, on their way north, he asked Ed to stay on. Apparently, Ed was impressed with the area and what Mr. Stoddard was doing and thought that was the right thing for him to do. So Mr. Stoddard hired Ed as his assistant, and he started work in July 1934. They spent the next decade running the CQSA.

Unfortunately, there was not a job for both Roy and Ed, so Roy took a position with the North Carolina game and fish department. He stayed there until an opportunity came for him to come to Thomasville, which came about during the cattle fever–deer tick episode in Florida. The Florida cattle industry, along with the help of the U.S. Department of Agriculture (USDA), was ready to wipe out the deer herd on the Seminole Indian Reservation at Immokalee because they thought the deer carried a tropical tick that transmitted a fever to cattle. It was so bad that they put up a cattle quarantine on the Georgia-Florida line. Every road had a dipping vat to prevent the transport of an infected cow over the state line. During the 1940s, when this particular episode was coming to a head, the Audubon Society was one of the few watchdog groups that would check up on the government, and they contracted with Roy—with Mr. Stoddard and Ed as supervisors—to go down there and conduct a study to see what was really going on. I know the Audubon Society was concerned about killing all the deer, as were Roy, Ed, and Mr. Stoddard. If they could, they were going to put a stop to it. So Roy resigned his job in North Carolina and came here to prepare for the study. Roy lived on the Seminole reservation at Immokalee for a year in a chickee—the traditional Seminole house—collecting and processing deer to see if they carried the ticks in question. He showed that they did not, but his study was not necessarily decisive from a policy standpoint. The USDA was going to stop the eradication anyway, as it was causing too much controversy. Anyway, that is how Roy found himself in this part of the world.

While Roy was working on that project in the early 1940s, Ed started a farm and game consulting business. Just as the quail plantations did with timber, they felt like they needed to contribute to the war effort on the agricultural side of things, so they started expanding their farming operations. Ed studied up on farming—he became a cattle farmer himself—and consulted with most of the quail plantations in the Thomasville-Tallahassee area. After the war, Greenwood Plantation hired Ed to manage the place, and he soon got Roy there, too. From that point until the founding of Tall Timbers, they were associated with Greenwood more than anything else. In some ways, it was Ed's departure that opened a slot for me with Mr. Stoddard. He did not find another full-time assistant until I came along.

Along about the late 1930s, while the second quail study was ongoing, Mr. Stoddard saw an opportunity to take advantage of his position as an expert in game lands. The hiring of Ed Komarek in 1933 allowed him to spend more time advising in other regions of the South, and he soon became deeply involved in the establishment of new hunting preserves. As early as 1928 he was already fielding requests from wealthy northerners to identify available property, and on his consulting trips he was already assessing the quality of lands for prospective buyers. But the organizational structure of the CQSA did not allow him to directly engage the real estate market. This changed in 1936, when the membership voted to release a portion of his time to form a real estate firm specializing in hunting preserves. Mr. Stoddard teamed up with Richard Tift—the half-brother of my former UGA classmate, Robert Tift—in Albany, Georgia, and Jack Jenkins in Charleston, South Carolina, both long-time real estate men in their respective regions, to form the firm Stoddard, Jenkins, and Tift. Mr. Stoddard had met Mr. Tift not long after he moved to Thomasville. Mr. Tift was an aggressive, successful man in the real estate business up in Albany, even in the 1930s when that was a tough business to succeed in. He was putting together some of the farmlands up there and making quail plantations out of them. Mr. Stoddard performed the on-site land surveys, evaluating the environmental conditions on particular tracts of land and making judgments on converting the land into hunting preserves, while Mr. Jenkins and Mr. Tift handled the business end of the operation. As Mr. Stoddard saw it, this arrangement gave him an even more effective role in implementing conservation in the South. More than that, this business widened his opportunities to participate in the forestry activities of the preserves, thus making it crucial to the emergence of the Stoddard-Neel Approach.

It is interesting how they put these large blocks of land together. In the early days you had to have a sizable unit for a quail preserve. The bigger it was, the better it was, usually. There was a limit on what was available, but Richard, being in the real estate business, worked on consolidating large properties. He would keep his eye on particular tracts of land for a few years before he was able to do anything, but then when the opportunity came he would buy two or three places that joined and create a preserve. The economic conditions of the time were pretty important for something like that to come together. During the Depression, the land was comparatively cheap, and wealthy sport hunters began snapping it up. As for the local landowners, the economic conditions were such that a lot of people had to sell their land. So in some ways the Great Depression was a critical moment that allowed for these properties to be cobbled together.

A lot of the plantations put together by Stoddard, Tift, and Jenkins still exist up in the Albany region, but ecologically they are not like they once were. Ichauway Plantation, which is now the home of the Joseph W. Jones Ecological Research Center, is a prominent exception. Mr. Tift helped put Ichauway together for Robert Woodruff, then the president of Coca-Cola, and Mr. Stoddard consulted up there as part of the Cooperative Quail Study Association. Mr. Stoddard wrote several early reports on Ichauway and advised on wildlife management in the early years, but Mr. Woodruff did not hire us for his forestry work. I first met Mr. Woodruff in the very early 1950s with Mr. Stoddard and Mr. Tift. Mr. Stoddard and Mr. Woodruff were friends, and Mr. Woodruff followed all the quail guidelines already established by Mr. Stoddard. Ichauway is still in existence as a viable, operational property that is maintaining and improving that wonderful longleaf-wiregrass ecosystem up there. It is not dedicated entirely to quail anymore, but a portion of it is. I am glad it is there because most of the other plantations we worked on are in terrible shape. Too much timber has been taken off of them in recent years. The Ichauway Plantation land base not only houses the Jones Center, but the entire property is the field laboratory for the detailed study of the longleaf-wiregrass and associated ecosystems.

I also worked on Pineland Plantation in the Albany area, which General Richard King Mellon owned. General Mellon was from

Pittsburgh and one of the heirs to the original Mellon fortune. He took over the family's business in banking and finance in the 1930s and eventually became an important philanthropist, especially around the Pittsburgh area. He was protective of his land down here, which is the kind of landowner I like.

I have an interesting story about General Mellon, one that illustrates how conscientious and concerned the landowners of that period were. When I was marking timber on his property in the 1960s, General Mellon would occasionally follow me through the woods watching from a distance—like a lot of the best owners did—to be sure I did not mess up his woods. He had a lot of young slash pine that Richard Tift had planted in what once had been tremendous peanut fields up there. He had planted those trees years earlier in order to break up the fields for quail habitat. I put the first thinning through these stands back in the early 1950s. Richard Tift had to sell General Mellon on the idea of cutting anything, but Richard knew enough about quail management and land management to know what was right. I would go out and mark a stand of planted pine pulpwood, and then Richard would take General Mellon out there to look over the marking. Finally, General Mellon wanted me to go with them one day. I went over there and we walked the woods a little while, and General Mellon said, "Why did you mark that tree?" I explained it to him, and he said, "Well, why did you mark this tree?" And I explained it to him. He continued on, "Well, then, why did you leave that tree," and I explained that one compared to this one. He wanted to know about every tree, why I either took it or left it.

Then we got to a tree with a ground line fork. Such forks can be genetic, or sometimes with planted pines you plant two seedlings in the same hole and you get two stems coming up. This happened to be one of the latter, and I had marked it because we generally took the forks out. He said, "Why did you mark that one?" And I said, "General Mellon, that is a forked tree. Both of them look healthy, but over the years pine straw will collect in the fork at the ground line, eventually a fire scar will develop, and those trees will be severely damaged. The growth will not be the same, and we are just releasing the ones on either side that are not forked." He said, "Well, think about this, if we had two trees everywhere that we've only got one, we'd have twice as much timber as we've got now." I had to say, "Well, yes sir, I guess you're right." The point is, even if General Mellon was not deeply informed about marking timber and other forestry practices, he was an engaged and concerned landowner. George M. Humphrey, who owned Milestone Plantation and was the Secretary of the Treasury under President Eisenhower during the 1950s, did the same sort of thing.

The last major plantation Tift and Stoddard put together in the Albany region was Nilo. Most of it had been a cattle farm, and Richard sold it to John Olin, the president of Olin Industries and the former head of SAAMI. "Nilo" is Olin spelled backwards. Mr. Stoddard and Richard Tift developed it into a quail preserve, and I was working with them then, so I did a tremendous amount of work on Nilo. It had to be restored from a heavily grazed cattle ranch to a top-notch shooting preserve. We worked hard on that and had a lot of support from Mr. Olin, who was dedicated to developing that land. The work we did at Nilo was a great example of how, through fire management and careful forestry, even land that has been extensively transformed by agriculture can be restored in ways conducive to the production of wildlife.

During and after World War II, the quail preserves experienced tremendous change. Modern, mechanized agriculture finally arrived on some of the preserves. Such developments not only displaced a large number of the tenants from the land, but they also altered the ecology of the quail plantations and other quail producing lands in ways that eliminated some of the unintentional ecological virtues of the patchy agriculture practiced under the tenant system, and added some of the environmental costs that came with mechanization.

There were occasional benefits that sprang from these agricultural changes. I can remember, for instance, when the first corn combine came into the quail preserve area in Thomasville,

Georgia. That was a modern machine, and it made harvesting the corn much easier and faster, but it also wasted a lot of corn. Grains of corn were knocked off in the process of being combined, so it left a tremendous amount of corn in the field. On the quail preserves in the Thomasville region, we had an increase in the quail population over the next several years due primarily to the fact that combines were dumping bushels of corn on the ground wherever there was a cornfield. The corn combine, then, was a good thing for the quail population in the short term. But it had a disruptive effect in the long run. A combine is a big piece of equipment. You can run a combine over a hundred-acre field of corn a lot cheaper per acre than you can run it over a two-acre field. To make it efficient you have to expand the fields, and that is what happened in a lot of farming areas. Field sizes increased, and patch farming simply could not keep up. That was not necessarily a bad thing from an agricultural standpoint; experts could boast about increasing yields, and those farmers who could afford to farm on such a scale benefited tremendously. But when they expanded the size of fields we lost a great deal of wildlife habitat and biodiversity. You also saw a lot of knowledgeable people move off the land.

Mr. Stoddard hated to see the expansion of commercial agriculture after World War II because it immediately depended on and enforced bigger fields. Some farmers even cleared valuable timber to put in big fields. Down here in the Red Hills, that was a double blow as far as Mr. Stoddard was concerned. They were expanding fields and making new ones at the same time. You clear a hill and put it in agriculture for a year or two, and then you have to abandon it because all your topsoil is gone. He thought that was just a stupid thing to do. Then they had to re-create a stand of timber and coax back all of the understory diversity. It was a waste.

Mr. Stoddard was also disappointed about the increased use of chemicals, especially regarding the government's fire ant eradication campaigns. He was an ornithologist first and foremost, and when the USDA started dumping millions of dollars worth of chemicals across the southern countryside in the late 1950s and early 1960s, and killing hundreds of thousands of songbirds and other wild creatures, he was very upset. He had been working on localized control of fire ants for many years, as he had some concerns about fire ant predation on quail, but he would never have advised the broadcast spraying of thousands of acres. That was just reckless. He could see the needless destruction and ecological damage that were inevitable with such a system.

We had meetings about the issue all over the quail preserve area, including several meetings up at Radium Springs, Georgia, for the Albany and Thomasville regions. The USDA circulated a lot of propaganda about how the spraying would not cause any harm, and, of course, most farmers supported whatever it took to achieve eradication. Mr. Stoddard and Richard Tift were among the few who were urging caution at these meetings. The first year or so after fire ants are established, it is hard to kill all of them anyway, even with powerful chemicals. The USDA spraying program never worked as far as eradicating them—there are just as many around here now as there were then. And they dumped millions of dollars worth of the strongest poisons like heptachlor, dieldrin, and mirex all across the South. We started organizing our own meetings—Richard Tift at Albany would promote them up there, and down here we would put them on at Tall Timbers. There would always be both sides represented. Many a time you would get an old farmer in there who would say, "Hell, y'all don't know what you're talking about. They dumped that stuff on my fields and I've got just as many birds as I ever had." That was a difficult claim to disprove, even though Mr. Stoddard was picking up dead birds all over the countryside. That did not make any difference to some. They just kept on insisting that they had "just as many birds as they ever had."

Mr. Stoddard was a practical man. He knew that people needed to make a living from the land, and that meant doing what you could to keep up. But a blatant disregard for nature was just beyond him. He would never advocate a mass attempt to liquidate any species, even if it was an invasive pest. The boll weevil was another great example. The boll weevil invasion was in full swing when Mr. Stoddard conducted the original quail study, and he wrote about it a good deal in *The Bobwhite Quail*. He came down particularly hard on agricultural experts who told farmers to clean up their fencerows and field edges, which were the prime habitats for the overwintering of boll weevils. Of course, when farmers cleared out all of that great edge habitat, they eliminated a lot of quail cover. He saw the situation a little differently. He did not write this, but in his mind I bet that he would say, "Well, the boll weevil would not be there if the cotton was not there, and the cotton does not belong there anyway, so if you want to get rid of the boll weevil, then get rid of the cotton." If you mono-crop a swath of land that stretches across the entire Southeast, you are setting yourself up for trouble. I have heard it called a buffet for the boll weevil, and that is exactly what it was. It was this type of land transformation that revealed to Mr. Stoddard what was at stake in the practice of ecological forestry and land management. Mr. Stoddard would laugh it off if he heard me say it, but I think that was what he rebelled against. He was not opposed to the cotton per se, but he was saying that he would just as soon have the brush and the boll weevil, and the hell with the cotton. There were other ways to make a living from the land.

The immediate postwar years also saw the expansion of the pulp and paper industry throughout the Southeast. Most of the region had been pretty well cut over between 1880 and 1920, which was when the longleaf forest really became imperiled. Of course, the Red Hills were, for a variety of reasons, one of the few areas where nice stands of longleaf had been spared. But the pulp and paper cutting was a new round of timber harvesting, and they often took trees to grind into pulp that would not have been marketable before. When the industry began to expand, we had several of the bigger companies move into Georgia and Florida. Georgia Pacific, Weyerhaeuser, and International came down here. We had one paper company, which shall remain nameless, that sent a team of their executives into the Thomasville-Tallahassee area in the 1950s. They wanted to build a paper mill, and they contacted Mr. Stoddard because they knew he controlled a good bit of timber. Mr. Stoddard called me one day and said, "I have some people coming by my house this afternoon. Come on over and we will talk to them about the paper business." So I went over there. They had sent the grandson of the president of the company, their chief accountant, and one of their foresters. They came in and gave what I considered to be a good old Chamber of Commerce talk. They said, "Now Mr. Stoddard, I know you will be real pleased with this. We are going to build a paper mill." Mr. Stoddard did not say much to them at the time, but he would have fought a paper mill tooth and nail, if nothing else because of the stench from the mill. He was not interested in having a paper mill close by, in our beautiful area, and none of his clients were either, not a single one of them. The executive went on with his speech, saying, "We have some choices. We can build a mill in Albany, or we can build it in Thomasville, or we can build it in Valdosta. Now, what we want to do, because you are a fine fellow, we want to give you the opportunity to do something for your community. With your help we will build it here." Mr. Stoddard said, "What kind of help have I got to give you?" They had done their research, and they knew Mr. Stoddard influenced five hundred million board feet of timber in the area. They said, "You give us the right of refusal of all the timber you sell, that is all we ask, just the refusal, and we will build the mill here." Mr. Stoddard would not have any of that. He said, "I can't do that. You all are wasting your time. We are not interested in getting the refusal from you on our timber. We believe in putting our timber up for bid, and that is the way we are going to continue doing it." And that was that. They ended up building the mill somewhere over in southeast Georgia, which was just fine by us.

Another big company opened up an office in Thomasville a few years later, though. They called it their regional forestry office. One day I got a call from the secretary of one of their officials. She said, "Mr. Neel, I believe you are a forester, is that right?" I said "Yes, ma'am." She said, "Well, Mr. So-and-So would like for you to come by his office. He wants to talk to you." I said, "Who is Mr. So-and-So?" "Oh, he is our regional forester." I was tempted to tell her he could come by my office if he wanted to see me, but then I reconsidered and told her I would drive by there. Well, I got over there and he had this intricate map of all the hunting preserves in the Thomasville-Tallahassee area. The plantations have their own field trial club called the Georgia-Florida Field Trial Club, and they had put out a private map showing all the members and their land. There are copies around, but the plantation owners have always protected that map pretty closely. Somehow, this fellow had acquired a copy of this map, and it was hanging on the wall right behind his desk.

So he said, "I understand you sell some timber every now and then. Well, I want to buy your timber, and we are going to offer our services. I have never seen timber in such bad shape in my life, all this old timber rotting on the stump. We need to get some good forestry in here." He got up and went back to his map and said, "You see this map?" Well, of course I had, because I helped to keep it up to date. I am sure I had a pretty cold tone by this point. I said, "Yeah, I see it." He said, "This map shows all the plantations here, and they all have a lot of timber on them." I said, "Is that right?" He continued, "And I'm going to buy every damn bit of it." I said, "Well, good luck with that," and I left. They were gone in a year and half. There was no way they were buying any timber around here talking like that. So the big timber and paper companies really made an effort to get into this area, and they ultimately did, of course, but we resisted them as much as we could.

Some of my most memorable times with Mr. Stoddard happened outside of the context of our forestry and land management work. Mr. Stoddard's interests were by no means limited to quail preserve management. He continued to consult for years with the Biological Survey on various wildlife projects. He was also the primary contact in the Southeast for a number of national conservation and ornithological organizations, and they often came to him with small projects they needed completed. As Mr. Stoddard got older, and as he gained more confidence in me, he included me on these projects as well.

We participated in the trans-Gulf migration study in the early 1950s, at a time when the details of bird migration were still largely unknown. I am sure people had ideas, but nobody had ever determined whether birds flew across the Gulf of Mexico to get to their wintering grounds in the Yucatan and Central America and on south. A lot of people thought the birds went around the Gulf and followed the coastline. So Dr. George Lowery, an ornithologist at Louisiana State University, instigated a trans-Gulf migration study that involved observing the disk of the moon with a telescope for three nights a month—one day prior to the full moon, the day of the full moon, and one day after. That gave us the most viewing time. We did it three months in the fall during heavy migration, and three months during the spring migration. Dr. Lowery got a series of observers to work with him from Florida around the Gulf coast all the way into Mexico and on to the Yucatan. He called Mr. Stoddard to see if he would set up an observation station along our Gulf coast. He tried to get them every hundred miles or so, I think. Mr. Stoddard jumped on that like he had not had a thing to do for a hundred years. It was just the sort of project he loved.

When we went out to observe, we were gone for two or three nights and altogether the better part of three or four days. Julie would fix us something to eat; she would hard-boil about a dozen eggs, and we would get some canned sardines, some canned deviled ham, some fruit, and water. We would go down to the Gulf and work from Alligator Point all the way around to Apalachicola. When we got past Alligator Point, we observed all the way up and down St. George Island. There was nothing on St. George Island but the lighthouse, the keeper's house, and cattle. Nobody lived over there. The man who looked after the cattle came over by boat from Apalachicola occasionally. So it was a wild place, and the only way you got there was by boat. We set up that telescope and watched all night long, recording the birds we saw. We each took thirty minutes on the scope and thirty minutes off, or an hour on and an hour off. You could not watch for much more than an hour without bombing out. Somebody had to record anyway, so the person on the scope would call out the data as it occurred and the other would record it.

The basic idea was to look through the telescope and count the birds that crossed the disk of the moon. We had the telescope lens set as a clock face—Dr. Lowery had all this worked out as protocol for the study. We recorded the actual time to the nearest minute when birds crossed. Then if a bird came in on the left and went out the right, it would be coming in at 9:00 and be going out at 3:00. There were three categories of clarity that he had worked out to estimate the height of flight. If it was real close it would be blurry—you could see the difference. If it was in focus, with a 20x-power scope, that was a pretty set altitude. And if it was higher than that it was some other category you could separate. We had to record all that. We sat there on some nights when there were so many birds coming over that you could not record the data fast enough. We could record up to eight or nine per minute. There would be so many sometimes, though, that you could not count them all. You just could not catch up with them. The other extreme was when we would be there all night long and never see a bird.

These observation trips were interesting the first few times, but after a while they got quite dull. I was interested in the results, but I was not particularly interested in gathering the data. I was only about twenty-five years old, my beautiful wife was home by herself, and there I was out there on a barrier island with Mr. Stoddard counting birds. If I had to make a choice, I can tell you where I would rather have been. But Mr. Stoddard loved those sorts of projects; that was the kind of person he was. He was more engaged in that than he would have been hunting gold nuggets in a stream. He had that ability to pick out what he wanted to be involved in and then put his whole self into it. And that was true when he was marking timber and making land management decisions in general. That project, by the way, was tremendously successful in establishing the existence of trans-Gulf bird migration.

We also spent much of 1951 and 1952 looking for ivory-billed woodpeckers in the Apalachicola River and Chipola River swamps at the request of the National Audubon Society. That was extremely interesting, but it was hard work, too. We spent days and days in that swamp all the way from Blountstown south to the marshes above Apalachicola. Whitney Eastman, the vice president of General Mills and a nationally known birder, had reported spotting an ivorybill down there, which made the ornithological world very excited. John Baker was president of the National Audubon Society at the time, and, of course, Mr. Stoddard knew him. Because of his early years in Florida, Mr. Stoddard had seen more ivory bills in his lifetime than just about anybody alive then, so Baker contracted with him. I do not know if they ever paid him anything—knowing Mr. Stoddard, he probably did it for nothing. They asked him to study the swamp, to go in there and research it and see if he could determine if there were any ivorybills in there, which he jumped on like a mockingbird on a june bug. It just so happened that Roger Tory Peterson, the great American ornithologist, and James Fisher, an English naturalist, came to visit during this time, and they joined us on one of our trips to the Chipola River. They were in the midst of touring around North America, and they wrote about our search in their book *Wild America*.[5]

Mr. Stoddard always drove an Oldsmobile. He would buy Mrs. Stoddard a new Oldsmobile every few years, and then he would take her old one. So we were always in the woods in that Oldsmobile. And, of course, we had a canoe, a Grumman canoe we called "The Pearly Bill." We had a sidearm that bolted on the back with a little 2.5-horse outboard motor you could put on the canoe. He did not like the motor, because it made so much noise and we could not hear the birds, so we only used it to run to particular places and in emergencies. We usually did not even use it when paddling back to camp, upstream.

We went down to Scott's Ferry on the Chipola River, to the last available place that you could drive to on dry land, which was a place called Bone Bluff. A man living down there, named Muriel Kelso, had first contacted Eastman about the ivory-billed woodpecker. He had a little concrete-block house in the swamp. Mr. Kelso was a river rat and a Pentecostal Holiness preacher. When I say he was a river rat, I do not mean that disparagingly. He just spent all his time on the river and salvaged everything he could that came down the river past his place. His land bordered the river, and there was a little hardwood drain that came into the river and always had a little trickle of water coming in. And then beyond that was swamp, more or less. He took a dead cypress tree that was probably thirty feet long, an old heart cypress tree, and made a boom with it by chaining it to a big tree where it would not come out. Then he swung it out in the river and pulled it around to where it was at a slight angle to the river coming down, and anything that floated down that river got trapped in that boom area. Sometimes he would find a boat, or a paddle, no telling what. It was there twenty-four hours a day every day, and he would go down and see what he had in his trap when he got curious. He also had one or two little cabins that he rented out. He called it a fish camp. So we went down there, put the canoe in, and disappeared down the river. We worked that river for a couple of days at a time, daylight to dark, all day long. We did that, on and off, for about a year and half, but we never saw an ivorybill, or any sign of one.

But Mr. Stoddard did see three ivorybills during the time I knew him. He saw two females in 1952, over in Ward's Creek swamp on the Greenwood property known as the Mitchell-Swift place, which is right here in Thomas County. I was not with him that day. He went into that swamp, and there was a pine island that had spruce pine on it, big sawtimber-sized spruce pine. Somehow a bug kill had gotten started in there, and most of the trees were dead. Mr. Stoddard went in to mark that bug kill to salvage the timber. Regrettably, he had sent me somewhere else that day. I got home late in the afternoon, and he immediately came over and told me that he had seen two ivory-billed woodpeckers. He was marking in that bug kill, which covered a couple of acres, and he heard them coming. They came in and lit on a couple of trees, and he thought they were going to feed on the bug kill, but they stayed around just a little bit and then took off to the north.

We spent a good part of the next year in that swamp. We went there a lot of mornings before daylight, because the best time to hear ivorybills would be early in the morning when they came out of a roost hole. But we never saw another sign of them. He swore me to secrecy; he would not tell a soul, because, he said, "if I tell anybody, that swamp's going to be overrun with bird watchers looking for the ivorybill, and they will destroy it."

His second sighting was from the air. He was flying over the Altamaha River swamp a few years later with Richard Tift, and he saw another one. Richard's brother-in-law owned a flying service in Albany, and they were trying to sell a plantation up close to Aiken, South Carolina. Somewhere between here and Aiken they ran into a tremendous thunderstorm, and they had to detour around it. The pilot turned toward the coast, to get around the storm, and Mr. Stoddard lost where they were in relation to the ground. They were flying at about three hundred feet, and he saw this big dead cypress sticking up above the canopy of the swamp. He looked and there was an ivory-billed woodpecker above the canopy on that dead cypress top. As they approached, the bird flew off, so he not only saw it sitting, but he also saw it flying. We went back over there as soon as he got home from Aiken. We had to load that canoe and go over there and find that ivorybill. Well, the Altamaha River swamp covers a tremendous area, and he had no idea where he saw it; he could only guess. We went in there and it was hopeless, so we did not go back again.

Mr. Stoddard did not have a fatalistic attitude about disappearing nature, but there were some things he could not do anything about. He could not do anything about the ivory-billed woodpecker disappearing from the face of the earth. There have, of course, in recent years been some hopes raised that a few ivorybills may yet be out there, but there is no definitive evidence of that yet. Mr. Stoddard did not like to see the ivorybill disappear, and the woodpeckers he saw he would not report. So he just had to shrug it off. I think he had a sense of realism. It was just a situation he could not do anything about on lands that he did not control.

But Mr. Stoddard did worry about disappearing nature on the lands where he did have some control, and we talked about that a lot in the years we worked together. Indeed, the Stoddard-Neel Approach really matured as we came to understand what we were losing in these landscapes, what had already been lost, and what might be salvaged and restored. It started as a quail management technique, and after the war it expanded to include more timber production, but all the while we were gaining a deeper sense of the values out there that needed representing and protecting. Within the confines of the management plans and goals on each of the properties we managed, we worked hard to understand, protect, and enhance those ecological values.

Mr. Stoddard always let me know when he thought some land-handling decisions were the absolute wrong thing to do. But still, in the Red Hills, it was hard to do much about it if a landowner was set on a particular management decision or direction. This is a privately owned landscape, unlike most other conservation landscapes, and the owners have a great deal of leeway in what they can do with their property. All you can do is to try to convince them that saving what they have is worth the effort, and that we might be able to help them meet their management objectives without sacrificing ecological values. The promise of the Stoddard-Neel Approach, of course, is that landowners can have good hunting, harvestable timber, and even some agricultural production if need be, and yet they can still have a healthy ecosystem out there as well. It is a conservative method and it requires some sacrifice, but it is not a matter of banishing all economic use from these landscapes. It may not be the most profitable form of land management in the short term, but it can pay in the long term.

We also need to remember as well that the natural world is resilient. An ecosystem, whatever condition it is in, is not a fixed, stable thing, and it cannot be preserved or managed as such. It is changing all the time. Everything out there is alive and growing. Living things are being born and things die; it is just not static. Conservation must be exactly the same way. If you study Mr. Stoddard's bobwhite quail book and the distinguished land management career that flowed from it, you will see change, adaptation, and evolution. His landmark study, written in the late 1920s and early 1930s, was the first scientific investigation of management on the plantation lands since they came into being in the late nineteenth century. His land management approach was not fully formed at that time, though you can certainly see what a fine naturalist and ecological scientist he already was. His thinking would grow and change, of course, but so would the landscape around him. If you were to examine the same land today that he examined in the 1920s, you would be hard pressed to recognize it as the same. Look at the photographs in his book—you see an entirely different picture of the land than you would see today on the same acres. The natural environment is moving forward all the time, and with our Stoddard-Neel Approach we are trying to create optimal conditions in their own time and place. It is a difficult thing to do, and yet that is the focus of our management.

Take the old-growth stand at Greenwood Plantation, the Big Woods, whose value, both economic and ecological, Mr. Stoddard did so much to protect and enhance. Some people call it a stand of virgin timber, but that is debatable, depending on how you define virgin. We know that it is a multiaged class of timber with an old-growth component. Some of the longleaf pine trees out there are four hundred years old or more. And the ground cover holds an incredible amount of diversity because of the simple management practice of regular burning. But there are also a lot of off-site species out there that make it a different place than it was fifty or one hundred years ago. And we have taken a lot of timber off Greenwood. The point is that nothing is entirely stable, and our conservation management cannot be either. That was one of the most important lessons that I learned in my years working with Herbert Stoddard.

The other critical lesson I learned from him was that, to manage a piece of land well, you have to pay attention to it and bring a broad set of interests to your management. One modern trend that hurts conservation, I think, is that people are getting further away from the land, which makes it harder to understand what is going on out there. For many people, even some conservationists, their understanding of the environment now comes primarily from reading about it, or keeping their eyes glued to a computer. They are not getting their hands in that dirt year after year to see what happens. On the other hand, we know more about particular ecosystems than we ever have because of the important work of ecological scientists. We have the potential to translate that knowledge into substantial conservation, which was exactly what Mr. Stoddard was trying to do all those years. But to do that well, we will need a new generation of land managers who are willing to combine scientific training with a commitment to gaining hard-earned experience in the field. We will need more people like Herbert Stoddard.

CHAPTER 3

The Early Years of Tall Timbers Research Station

During the 1950s, after I came to work for Mr. Stoddard, there slowly emerged a conversation about creating an institution for scientific research that would carry on the work in the Red Hills that had been initiated by the Quail Investigation in the 1920s. Mr. Stoddard was obviously the key to that, as so much of the management knowledge we were working with had resulted from his careful experiments and observations. But with Ed and Roy Komarek, we also had two other good scientific minds to add to the mix. The idea was to find a way to house and fund scientific research on the longleaf system that could continue to inform conservation and practical land management. The eventual result was Tall Timbers Research Station, founded in 1958.

To really understand how Tall Timbers Research Station came to be, you first have to understand the community from which it was created. As a research station, Tall Timbers materialized through a great deal of intentional thought and purpose, but it was also an organic institution that sprang from an extraordinary community of naturalists living in the area. Nobody was in this for his own personal gain or for scientific notoriety; it was done out of sincere curiosity and concern. Henry Beadel, Herbert Stoddard, and Ed Komarek founded Tall Timbers to study and perpetuate the longleaf-grassland plantation landscape they had been working on for years and years. While it is rarely mentioned in the official histories of Tall Timbers, Roy Komarek and I were founders as well.

Living at Sherwood gave Julie and me access to Mr. Stoddard's friends in the area. Mr. Stoddard's closest personal friend was Henry Beadel, who lived at Tall Timbers Plantation just across the Georgia-Florida line. Mr. Beadel's uncle purchased Tall Timbers in the 1890s, so he had been around the Red Hills area for quite a while. He started coming down every winter as a younger man, and he bought the place from his uncle in 1919. Mr. Beadel was born to a wealthy family in New York, took a degree from Columbia, and was a practicing architect for many years. He loved Tall Timbers, and, after World War II, he decided to take up full-time residence down here. He and Mr. Stoddard were about the same age, and they hit it off from their initial acquaintance at the start of the quail study. Mr. Beadel differed from the other preserve owners in that he was a naturalist first and a hunter second. He was an excellent amateur ornithologist and photographer, and Mr. Stoddard recognized that from the beginning. He even commented in his Quail Investigation field diaries that Mr. Beadel and his wife were "both intensely interested in Natural History and he is really quite a naturalist, having a knowledge of the majority of the birds and reptiles of the vicinity." Mr. Stoddard also appreciated that Mr. Beadel had "quite different ideas in regard to hawks and owls and will not allow any to be killed on his place."[1] It is hard

FIGURE 6. Herbert Stoddard, left, and Ed Komarek, right, pursued wildlife research and land management together for almost four decades in the Red Hills. Stoddard hired Komarek as an assistant in 1934, and Komarek eventually became a leading national authority on fire ecology as the longtime director of Tall Timbers Research Station.

to overemphasize how rare that was among the preserve owners in those days. They were good people, but few had the knowledge that Mr. Beadel did.

The Komareks—Ed, his wife, Betty, and Roy—had purchased Birdsong Plantation in 1938. Birdsong bordered Sherwood, and a family named Dickey had owned it since before the Civil War. It had been a working farm for generations; the Dickeys scratched out a living on that place for a long time. Most of it was in various stages of old-field succession by the time the Komareks bought it, and Ed eventually converted some of it to pasture. Ed and

Betty lived in the old Dickey house, and Roy initially rented an apartment in town, because there was not yet a house for him on Birdsong; they fixed up a house for him later on. But he was out there all the time.

We were in a rural area. There were no paved roads out there, and there were very few people around. Most of the adjoining property was undeveloped. Mistletoe Plantation was just down the road to the southwest, on the Florida line. Roswell King owned it in those days, and at the time it was just a few thousand acres of hunting land; it was not a developed quail plantation like it is today. Heyward Mason, who grew up on Susina Plantation just down the road, had married Mr. King's daughter, Edith, and they lived at Mistletoe. There was also a local family, the Browns, who owned a thousand acres or so next to Mr. Stoddard on the east. Dr. Brown was a country doctor who had built a practice for people in the immediate area. Beyond that was Susina Plantation. For about six or seven miles from the Florida line to the Cairo Road, there were just a few landowners, so everybody knew everybody pretty well. There was not much reason for anybody else to come out there, so we did not get much traffic on the road. It was just us and this spectacular landscape.

Mr. Beadel would often invite Mr. Stoddard and his wife to have dinner. Mr. Beadel's second wife, Beatrice, was still alive at that time. The Stoddards did not mingle in the plantation social scene; their friendship with the Beadels was based on companionship, mutual interest, and respect. Mrs. Stoddard, who was an excellent pianist and seamstress but did not share her husband's interest in the natural world, was close with Mr. Beadel's first wife, Genevieve. The Komareks were part of this circle too. Ed had worked with Mr. Stoddard since 1933, and they were still involved together on a lot of projects. From the time Julie and I moved out there in 1950, Ed was working for Greenwood Plantation as an assistant manager. Major Louis Beard was the manager of the Greenwood properties back then. Mr. Stoddard and Ed had a close friendship, because they had been involved together in the Quail Study Association.

After the association dissolved in 1943, they continued to work closely together. When Mr. Stoddard went into the real estate and forestry business, Ed started a consulting business in game and agricultural management for the quail preserves. Ed did that independently until Greenwood hired him full time, which was right around the time we arrived back from Athens.

While living at Sherwood, Julie and I got to know Mr. Beadel well. He had no children, and, not long after we moved there, his second wife died. Mr. Beadel did have a brother, Gerald, who had a small interest in the Tall Timbers land, but he did not come down very often, if at all. We saw Mr. Beadel frequently. Mr. Stoddard would go over there, and often he would take me with him. Sometimes, on special invitation, Julie and I would go over to have dinner at Mr. Beadel's. Julie even entertained Mr. Beadel at our little tenant house on Mr. Stoddard's land a few times, which, in retrospect, really amazes me given how modest the place was. Then Mr. Stoddard and I started taking Mr. Beadel on some of our outings on the Gulf coast. He liked to go with us to Bird Island in the bay at Shell Point, Florida. We would lead him to a blind, and then we would back off in the boat about a quarter mile out in the bay, waiting for him while he spent a couple of hours photographing birds as they came back to their nests. He was an accomplished wildlife photographer. I did not know many plantation owners, or people of the social status or wealth that Mr. Beadel represented, so he made a strong impression on me. He was a distinguished, well-educated man, and he was accomplished in all of the social graces. I, on the other hand, was not raised in exactly that same atmosphere. So I held him in awe.

As Mr. Beadel grew older, and in the absence of any obvious heirs, he began to think about the future of his land. He loved Tall Timbers and really paid attention to his land: he nurtured it, understood it, and took care of it. When I knew him, he spent most of his days out on that property. He would put up a blind in a chufa patch and photograph wild turkeys, especially in the spring when the gobblers were strutting. He would photograph anything that he saw that he felt was worth photographing. He was a hunter as well as a naturalist, but I think, if he was given a choice, he would rather carry a camera into the woods than a gun. Mr. Beadel had canoe trails dug through the shallows of the pond at Tall Timbers so he could photograph the bird life. They went through the marshy areas that in normal times had just a little water. They would not have supported a canoe, so he had channels dug where he could ease through there and photograph birds, or tie up and wait to see what he could see. He spent many happy hours out there.

He often asked Mr. Stoddard for advice about the future management and ownership of his land. Of course, they had naturalist friends from around the country who also gave expert advice on it, but Mr. Stoddard was an important confidant. Another person who advised Mr. Beadel was Richard Pough. Pough was a distinguished naturalist who worked with the Audubon Society and the American Museum of Natural History in the 1940s, and he helped to found the Nature Conservancy, becoming its first president in 1954. At Mr. Stoddard's suggestion, he came down to give advice on the future of Tall Timbers, and he acted as a consultant when Mr. Beadel was setting up the research station. Pough recognized the need for an apparatus to preserve private land throughout the nation, which was and remains the Nature Conservancy's primary aim. And, at the time, he was the authority on doing so. He was an ideal person for Mr. Beadel to be talking to.

The creation of Tall Timbers resulted from a long period of thought, discussion, and examination of what other people were doing, both in the region and in other parts of the country. Mr. Beadel's priority was always nature broadly defined—he was more interested in the diversity of plant and animal life that existed in the Red Hills than he was in managing land for one particular species such as quail. Hunting, or game animals in particular, were an interest, of course, but never his sole interest. He recognized this hunting preserve environment of the Red Hills as an important biological refuge, and he hoped that his property would be a model for the intentional management of the rest of the preserves. At the time, the quail plantations were still in excellent ecological condition, largely due to the influence of Herbert Stoddard.

While Mr. Beadel was thinking about what to do with his property, another important project developed near Tall Timbers that helped to make clear the potential value of his land for scientific research. Mr. John Phipps, who owned Ayavalla Plantation and was a close friend of both Mr. Beadel and Mr. Stoddard, approached Mr. Beadel about the possibility of building a television tower on Tall Timbers. Mr. Phipps, whom they always called Ben, had the first television broadcasting rights in the Thomasville-Tallahassee area. Tall Timbers was located just about exactly between Thomasville and Tallahassee on one of the tallest hills, which made it an ideal location for such a transmitting tower, which is why Mr. Phipps went to Mr. Beadel.

Mr. Beadel's first reaction was, "Certainly not! I wouldn't have that ugly thing on my place. I don't want anything like that." Then he talked to Mr. Stoddard about it. Mr. Beadel said, "Can you imagine Ben Phipps wanting to build a television tower on my property?" Mr. Stoddard said, "Wait a minute now, we better think about this a little bit." Mr. Stoddard was a conservationist who appreciated natural beauty and aesthetics as much as anybody, so he was reluctant to see such a tower go up. But he was also a realist. Television was a relatively new technology back then, and these towers were critical to broadcasting strong signals. They were going to go up throughout the area no matter what. More importantly, he saw a scientific opportunity. Mr. Stoddard had read about, and been amazed by, how these towers in other locations had produced tremendous bird mortality. Migrating birds were flying into the towers and being killed by the impact. As he understood it, there could be several thousand birds killed in one night in some cases. When Mr. Stoddard told Mr. Beadel about bird kills at other towers, bells started going off. They both sensed an opportunity. That tower was going to be built somewhere, and it was going to kill birds, so why not have it on Tall Timbers where Mr. Stoddard would have access to picking them up and studying them. In the end, Mr. Stoddard convinced Mr. Beadel that it would be an opportunity to study bird mortality that they should not pass up. Mr. Beadel changed his mind, and the tower went up in 1955.

Mr. Stoddard devoted much of his time during the next ten years to studying the grounds of that tower, and in the end he recorded well over thirty thousand individual birds of 170 different species killed by that tower. My friend Bobby Crawford wrote a great history of the tower study. He commented that "Stoddard's data made Leon County, Florida, the best-documented migration locality in the State, if not the region, with copious detail even about species notoriously difficult to find by ground observers."[2] Ornithologists still reference the data Mr. Stoddard gathered in that tower study.

Setting up the tower study took a great deal of hard work. There were the remnants of a tenant house and outbuildings on the hill that had been selected for the tower site. They had been there for a hundred years or more, I guess. The hill had fenced-off patches for hogs and an old cow barn. These were all sheds, and they were primitive in those days. All of it had been abandoned, but several generations' worth of accumulated debris remained. It had to be cleaned up, because it was within the area where the birds would fall. The site was also overgrown with all manner of weeds and shrubs. Mr. Stoddard was down there every day, and he saw what had to be done. So he came in and told me one day, "I want you to get Willie. We have to clean up around the tower down there." I said, "Yes, sir, we can do that. That's no problem. We can knock that out in a hurry." Willie Lurrie worked for Mr. Stoddard, and he was a fine fellow. He did most of the work on Mr. Stoddard's place that needed to be done. Willie and I went down there and got to work. We had to take up old fences and haul off a hundred years of trash, rusted tin, abandoned farm implements, and all sorts of stuff. We finally got it cleaned up, and we ended up clearing about forty acres around the tower. We did not clear all of that at one time, but we eventually had it looking like a golf course. We had to be able to find the birds on the ground, so we kept it close-cropped with a rotary mower. We mowed it at least once a week and sometimes more than that if we had a wet year.

Before the tower study, there had been some thought given to setting Tall Timbers up as a field station to conduct biological research. Of course, this was also an outgrowth of both the original quail study by the Biological Survey, which Mr. Stoddard headed, and the privately funded Cooperative Quail Study Association, which had continued Mr. Stoddard's work in the service of the hunting preserves and their management. Mr. Stoddard had continued to conduct research on how to best manage the longleaf-grassland environment, and to consult on a network of plantations throughout the South under these and other auspices. He was just an inveterate researcher and investigator, a quality that I think the tower study makes clear. He was always looking for opportunities to learn new things, and to confirm through experiments what he thought he already knew. With that experience, Mr. Stoddard envisioned all sorts of questions he could work on, especially in regard to fire. So when the tower study came about, Mr. Beadel realized fully what could be done. I think the tower study really convinced Mr. Beadel to devote the land to research. On top of that, here were two world-class scientific minds in Herbert Stoddard and Ed Komarek who did not have a research home, so Mr. Beadel really did us all a great service by providing them with one.

A lot of people assume, because of the emphasis on quail among the plantation owners, that Mr. Beadel established a quail research station, but that was not his primary purpose. He established a biological research station, and the first project was studying bird mortality at the television tower. Mr. Beadel was obviously interested in quail, but not to the exclusion of everything else. He had a hunger to know what was going on out there, how the system worked, and how to protect and enhance his land. That is why he found Mr. Stoddard so fascinating.

Tall Timbers has an interesting land-use history. It sits within the Red Hills near Tallahassee, on the shore of Lake Iamonia. Native Americans had a strong impact on the area around the lake, and while there are no known Spanish sites on Tall Timbers, there are some right across Lake Iamonia. So there were historical influences on that piece of property, perhaps more so than on some of the other places like up in Thomasville, for instance. Starting in the 1820s and 1830s, most of the land around Lake Iamonia was heavily farmed under the cotton plantation system, including Tall Timbers. There is a lot of old-field land there, which is visible today if you walk around the place. And as I mentioned earlier, there are some erosion gullies on Tall Timbers that you could hide a vehicle in. They are that deep and that steep, and were created by the early farming activities before good land stewardship. So Tall Timbers was not a pristine place. There was still quite a bit of tenant farming going on when the Beadels acquired the property, but many of the fields had been abandoned and had seeded into pine, loblolly and shortleaf primarily. So the first time I saw Tall Timbers in any detail it was a typical old-field stand of timber, relatively mature and burned regularly. They did some brush clearing by hand when it got out of control here and there, but basically fire was used to control the rough.

In 1941 we had a hurricane come through the Red Hills, and it blew down a sizable volume of timber on Tall Timbers. Mr. Stoddard was not yet in the forestry business, but this really started him in forestry, because a lot of his friends asked him to help them clean up their woods. It is amazing how a random event in nature can alter one's career trajectory. There were some interesting issues involved with his salvage effort. In those days they had the portable sawmills that they could move around and set up. They were not on wheels; these were very small sawmills, commonly referred to as peckerwood mills, that you had to anchor on timbers in the woods. So Mr. Stoddard got that operation going in 1941, on Tall Timbers and other plantation lands that he managed, and over the next two or three years this salvage effort led him to become deeply involved in the timber business. More than that, it gave him a first-hand look at a large-scale ecological disturbance and allowed him to experiment with landscape management and restoration after a substantial reduction in standing timber.

Mr. Stoddard used several mills to salvage the downed timber, but one in particular was a portable sawmill run by two brothers from Quincy, Florida, named Bill and Gus Mitchell. The Mitchell brothers grew up poor, but they were hard working. Mr. Stoddard got them to move over here, set up their mill from place to place, and saw the downed timber. After they cleaned up the damaged timber, they found a site in Metcalf and set up a permanent mill, which became Mitchell Brothers Lumber Company. They were fine men—honest, hard working, straightforward. The primary thing is that they were honest. They became one of the major buyers of Mr. Stoddard's timber, and then our timber. They made a tremendous contribution in working on the plantations, because they did an exceptionally good job in the woods. Such careful logging accounts for a great deal of the success of our, or really any, timber management system. Incidentally, when Mr. Stoddard's son, Sonny, got out of the army after World War II, he went to work for his father. Sonny was doing what I ended up doing until he bought a third of an interest in Mitchell brothers, and it became Mitchell Brothers–Stoddard Lumber Company. It was an excellent outfit until Gus was killed in an accident. Sonny and Bill ran it for a while, but they were both getting older, so they sold out. It has changed hands several times since then and has no resemblance now to the original mill.

That 1941 hurricane, and the careful forestry that followed in its wake, not only had an important impact on Tall Timbers plantation, but it is what really solidified Mr. Stoddard's consulting forestry business. Incidentally, this all occurred during and blossomed after World War II, which coincided with the end of Mr. Stoddard's privately funded wildlife research. Mr. Stoddard and Ed had phased out the Cooperative Quail Study Association during the war. Tall Timbers, then, was in some senses a rebirth of that old idea, though with a much stronger emphasis on forest ecology and timber management than on quail production.

I am not sure that Mr. Beadel would have used these words, but, in retrospect, it is fair to say that the original purpose of Tall Timbers was to study and perpetuate the ecology of the area. Most of the hunting landowners at the time generally shared that goal. In 1950 there were only one or two plantation owners who did not seem to value the beauty and the total ecology of their land. Even if the plantation landowners did not know much about the science behind it, they knew what looked good and had developed an informed aesthetic sense of their land. One or two of them, however, were more or less exploiters. They mistreated their timber, and they seemed to emphasize the greed of hunting more than anything else; they had the trophy mentality that Aldo Leopold criticized so often. But most of them—and this was a tremendous influence on me—shared Mr. Beadel's feelings for the land in one way or another. I had personal contact after a while with most of the owners, and I quickly learned that if I cut too many trees, or changed the look of the land in ways that they did not like, I was in trouble. They put aesthetics ahead of everything else, and that influenced me tremendously.

The atmosphere in the whole Red Hills plantation section was different then, and that lasted up until about the mid-1960s, when the older owners of my time began to pass away. The land was in most cases handed down to the next generation. Some of them were good conservationists, and some of them were not. There were also new economic pressures on this generation. Taxes were going up, and some new owners did not have the wealth to maintain these lands as their predecessors had. As a result, some of the lands were sold when the owners passed on. That was about the beginning of the end, when they sold. Some of them were sold to people who respected and loved and perpetuated the properties. But many others went to people who wanted to maximize both quail numbers and timber income, and, unfortunately, those two things went hand in hand, at least in the short term. You can cut a lot of timber and create good quail habitat through disturbance,

FIGURE 7. Sometimes large-scale forest disturbance is unavoidable, such as this tornado damage in 1971. In this stand and others—including Tall Timbers—Neel salvaged what timber he could, and then began the process of encouraging regeneration in the open areas. The Stoddard-Neel Approach recognizes that these disturbances sometimes do occur, and it deals with them as part of natural processes.

but that is not good long-term ecosystem management. You are not managing your land for the future.

Mr. Beadel was concerned not only with pure science, but also with doing science that would inform sound ecological land management. As far as his timber was concerned, the charter of Tall Timbers clearly indicated that there was not to be any timber cutting for commercial reasons. We did have to cut some timber once, however. A tornado touched down in 1971 that knocked over a lot of timber. I salvaged about a million and a half board feet off of Tall Timbers, most of which was big loblolly. I was on the staff and I handled the sales, but I could not charge anything for my services. Jimmy Atkinson, who is now the land manager for the Jones Center at Ichauway, was an apprentice manager at

Tall Timbers then, and we worked together on that salvage job. He monitored the cutting and was responsible for the quality of the job. But we had to get a legal exemption to salvage the timber once it had been blown down, because Mr. Beadel made it clear that he did not want any timber cut at Tall Timbers.

The 1950s was an exciting time to be in this isolated little corner of the region, because so much was going on of scientific importance. When Julie and I moved onto Mr. Stoddard's property we became part of that community. On Sundays, Mr. Stoddard had the habit of going up to Birdsong in the middle of the morning to have a cup of coffee with Ed, Betty, and Roy Komarek. And if anybody was in the area, a visitor with an interest in ecology or ornithology or land management, they always came as well. And there were a lot of people coming through to visit with Mr. Stoddard or the Komareks, all three of whom were well known throughout national conservationist circles. Mr. Beadel came fairly often, and Julie and I were there also.

Every Sunday morning was a seminar of sorts, and there was a world of knowledge gathered there to be shared with all. Ed Komarek was an important contributor to those seminars. He was energetic and an excellent field naturalist. Ed was like so many brilliant people: he did more than he recorded, and a lot of what he did was not systematized as modern science, because he did so many things at one time. He had studies of all sorts going on all over the place. From the time he arrived in the Red Hills he had accumulated data on the mammals of the region, most of which never got published. That mammal data has to be at Birdsong or at Tall Timbers, and someone ought to revisit it. He collected data on southeastern mammals for a long time, and he just never could quite finish it.

All sorts of people who were involved in natural history came through the area because they wanted to see Mr. Stoddard or the Komareks. So the "Sunday Morning Bunch," as we came to call it, went from an informal gathering over cups of coffee to a regular meeting of people in the area who were interested in natural

The Stoddard-Neel Approach

A Photo Essay

The Big Woods at Greenwood Plantation exemplifies the longleaf pine-grassland ecosystem at its best. This 1985 photograph shows a view representative of the classic "parklike" longleaf aesthetic that early visitors to the region would have seen throughout the Red Hills. It is a multiaged open forest that affords a view of considerable distance. Its openness allows adequate light to reach the forest floor—a low, grassy understory reminiscent of prairie land. By some accounts, a healthy longleaf forest with a native understory is the most biologically diverse environment in North America, and if managed correctly, it can be economically profitable over the long term. Although a longleaf-grassland system should be visually open, it should still carry a substantial amount of timber, depending on the site. Between 1945 and 1995, as part of our ecosystem management approach to the land, we harvested more than fifty-six million board feet from Greenwood while still managing to *increase* the standing timber by forty-eight million board feet. For over fifty years, the Big Woods was a favorite quail course of the Whitney family.

The Stoddard-Neel Approach advocates small gaps throughout the forest as the best way to encourage the sort of longleaf regeneration shown above. In nature, lightning strikes, blow-downs, and other disturbance events create these gaps, and we mimic such small-scale disturbances when we harvest timber. We prefer to catch the longleaf regeneration first and then release it gradually by increasing the gap size. When we see a heavy mast year, we modify our fire regime to burn just before

seed fall. Longleaf seeds need bare mineral soil to germinate, and a well-timed, low-intensity fire during a mast year creates perfect conditions to capture the seed fall. This type of management produces continuous regeneration in a multiaged forest, and it allows us to maintain the forest continuously. Hopefully, one or two of these trees will live to be quite old with management that perpetuates the total stand.

We always have to work with what we have available on the ground, and land-use history necessarily influences the decisions we make. Sometimes, as in the above case, we take over land that has been neglected, and we have to work hard on correcting past mistakes. This was a true longleaf-wiregrass site, but it had been heavily cut over and grazed in the past. Those are not perfect conditions to deal with when restoration is your goal, but the land still had wiregrass and was kept open with burning, which is more than can be said for many sites we have worked on. Here, we decided to regenerate longleaf naturally, and we were able to catch a tremendous amount of the seed fall. We altered our burning schedule to precede the seed fall and then skipped a year before the next burn, when we used a low-intensity fire to keep from harming the new seedlings. We have no standard formula in the Stoddard-Neel Approach. We always have options, and the choices we make depend on land-use history, available resources, and landowner goals.

In opening gaps for longleaf regeneration, you have to be careful that they are not too large. A large gap in the forest can create challenges with our burning regime, as is clearly illustrated in this photograph. We did not open this gap for regeneration—it was a wildlife food plot—but the same principle applies. The ground cover alone does not always provide enough fuel to carry a fire, especially on old-field land; we really need both ground cover and needle fall. In the absence of fine fuel such as needles, fire can skip over a large gap, leaving it susceptible to hardwood encroachment and a loss of both ground cover diversity and pine production. That is why maintaining a fairly consistent canopy structure is so important to the Stoddard-Neel Approach. Controlling gaps such as this one for future longleaf regeneration might require the extra step of going back to set a spot fire on a hot, dry day. Otherwise, such openings can revert to rough, and you will need to control them mechanically or chemically. Then again, we might want to create a quail covert in this gap. Again, we always have options on the ground. It is just up to the land manager to take advantage of those options for the good of the ecosystem as a whole.

This is a great photograph for illustrating the return of longleaf pine regeneration on old-field land. Compared to what you see in the Big Woods, the groundcover here is poor in terms of diversity, a reflection of depleted soil and an extensive disturbance history. We had a good mixed stand of longleaf and loblolly on this land, and you can see the ground cover was mostly broomsedge, or "poverty grass" as some southerners call it, which is a good indicator of former agricultural land. In a less patient approach to forestry, like even-aged management, you would have harvested all of the timber and then replanted rows of loblolly or slash pine in its place. What we did, instead, was to burn frequently while harvesting through the years with our method of single-tree selection, which continued the structure of the timber stand in a way compatible with fire

management. We patiently waited for a good longleaf mast year, and now we have the beginnings of a multiaged forest that will continue to grow into the future. Now we need to release these young trees by carefully opening gaps around the regeneration. We do not mind growing species like loblolly for the long-term goal of reestablishing longleaf, but over time we will select for longleaf by gradually removing the loblolly on the site while also maintaining a quality quail-hunting environment. The next step here is to reestablish a diverse understory plant community, which will hopefully come in time. There has been a recent move in the South to make native grass seed commercially available, which could speed up ground cover restoration.

Our management can accommodate all sorts of land uses. In this example, we have a three- or four-acre patch of corn in the middle of an excellent stand of longleaf. This is what much of the Red Hills looked like when Mr. Stoddard arrived back in the 1920s and when lots of tenant farmers were still on the land. From the air, you would see these little patches spread throughout the forest from property to property, with few absolute breaks in the forest canopy. Many aspects of tenantry are not present in this picture, of course. You see very little evidence of people actually living on the land. But the landscape itself is similar. There is less of this kind of

small-scale agriculture in the midst of good longleaf lands today because of intensified agricultural, forestry, and quail production, but we can still work with small fields in our management if landowners are interested in having them. Such small-scale agriculture, the creation of food plots, or small areas with early successional plant communities provide ideal food sources and edge habitat for wildlife of all sorts. But we want to be careful not to create new areas like these in places where there is diverse native ground cover.

This is a nice example of a managed quail covert with some wild azaleas in bloom on Pebble Hill Plantation. This is a great place for quail to hide from predators. We protect these coverts from fire during most years—notice the firebreak plowed around it—and only burn them every so often to prevent hardwood encroachment. Habitat like this will usually sprout back after an occasional fire. These coverts help boost quail populations in the forest, but, as in this example, we also get aesthetic byproducts, such as the wild azalea. One of Herbert Stoddard's pioneering contributions to the broader field of wildlife management was his advocacy for the creation and perpetuation of habitat diversity as a way of producing truly wild game populations. He shared this approach with his colleague and friend Aldo Leopold.

Here is the real biological heart of the longleaf pine forest—the ground cover. From a distance, you may just see a sea of grass and think the forest contains nothing but trees and wiregrass. Many an early traveler called the coastal plain uplands the "pine barrens" before they knew anything about this woodland environment. But we now recognize tremendous biological diversity out there, and most of that diversity is in the ground cover. The grasses, shrubs, legumes, and other woody plants provide food and cover for a host of wildlife species, and the trees offer both habitat and the resinous needles needed to carry fires. Fire, of course, holds this whole system together.

Along with fire, water is one of the most important factors in creating the diverse habitat gradients in the longleaf biome. Look at this beautiful wildflower bog. This is what we call an ecotone, a transition area between the upland forest and a small wetland drain. We cherish these areas because of the beauty and diversity they offer. Managing for a bog such as this one, however, is not as easy as simply deciding to protect it and then leaving it alone. These ecotones need to be burned every few years to stave off hardwood encroachment, a proposition that may test one's patience. There are only a handful of days each year when conditions are right to burn an area like this properly, and you can bet that it will be so hot and dry that you do not want anything else in the forest to burn. So you have to plan ahead for burning a bog. First, the uplands must be burned early to reduce the fuel load that surrounds the bog. Then you can burn the bog on a dry day without fear that the fire will escape.

Here are two of the best naturalists I know, my wife, Julie, and our dear friend, Angus Gholson, posing with our sparkleberry tree. Some desirable tree species like this sparkleberry would not survive in a fire-maintained forest without some help. This sparkleberry is not fire tolerant and it has no real economic value, so it would be easy to just let it burn or to cut it down to make more room for pine trees. But that would be a mistake, for it provides good wildlife food, and hairstreak butterflies love it when it blooms in April. For those reasons, we decided to make an effort to keep this tree. This particular tract is a loblolly stand with some hardwoods mixed in. We released the sparkleberry by clearing the close competition, and we always rake the fuels from around it to protect it from a fire. Not every land manager will go to such lengths to protect a sparkleberry tree. In fact, many probably do not understand its larger ecological importance. But the ecological land manager pays attention and takes advantage of these opportunities in nature when they present themselves. We saw an opportunity for more wildlife forage, and we took it. On top of that, this is a beautiful little scene in the middle of the forest. The Stoddard-Neel Approach makes a place for species like these.

Though we prefer to mark defective trees, we do not want to rid the forest of all of them. We actually have a high regard for many deformed trees under the Stoddard-Neel Approach. A witch's broom such as this one, for example, provides ideal habitat for many animal species, including owls, raccoons, and songbirds. Technically speaking, a witch's broom is a deformity, but it does very little harm to the tree itself. We call these "character trees," and they add a great deal of both biological and aesthetic diversity to the forest. We can keep an array of character trees such as this one because we select trees individually for harvest, and we keep a close eye on the logging process.

The Stoddard-Neel Approach to forestry does not end after we finish marking timber. We monitor all logging activity and demand that loggers adhere to very high standards, particularly when it comes to protecting ecological resources. In the above case, we marked off an area where a rare Bachman's sparrow was nesting. The Green family from Camilla, Georgia, logged this tract, and once they understood what we were trying to do, they took a real interest, as was often the case with our loggers. It is getting more and more difficult to find loggers who leave little trace of their activity, because of the heavy machinery they use today. We still have some loggers in the area who do a good job, but it is hard for them to maneuver their machinery without doing harm to other trees or the ground cover. Moreover, because of their investment in all of this heavy equipment, their cost of doing business is that much higher, which requires that they cut more volume than before to make ends meet. So, ecological forestry is difficult to practice in today's market. But if loggers want to cut our timber, they must adhere to our strict guidelines.

Herbert Stoddard started many fires in his day, literally and figuratively. He recognized fire's importance to the longleaf pine-grassland environment, he fought for its acceptance among foresters and land managers, and he used it prolifically. Even later in life—Mr. Stoddard was seventy-seven years old when I took this picture in 1966—he loved to go out and set a fire. Here he is on Sherwood, where he lived, setting a head fire that can take care of a rough in a hurry if done correctly. We used all types of controlled fire, depending on weather conditions and our goals. A head fire sweeps through a stand of timber with the wind, and it burns quickly and intensely. A backing fire creeps against the wind, burning more slowly and completely. Spot fires can be spread throughout the stand and are good when winds are variable. And we sometimes used flanking fires, or a combination of fire types depending on the weather and the experience of the burner. Mr. Stoddard always wore old pants that were frayed down around the cuff, and whenever he was out burning the fraying would usually light up a time or two during the day. It was my job to watch carefully and put him out when necessary.

This is what happens when fire is excluded from the longleaf forest for a long period of time. Roy Komarek took this photograph in 1964. For one reason or another, the landowners excluded fire from this stand for more than forty years. This kind of rough diminishes the longleaf forest from almost every point of view: aesthetically, recreationally, economically, and ecologically. You cannot see or do much of anything in a stand like this, and it is much less biologically diverse than an open, fire-maintained forest. Nonetheless, areas such as this one are ecologically interesting, and they often occurred historically in hammocks and other isolated portions of the landscape that were cut off from fire. In the Red Hills, when fire is excluded from a longleaf-wiregrass forest, it moves toward an oak-beech-magnolia upland hardwood climax. We can restore these areas to longleaf by eliminating the hardwood trees and reintroducing fire, but I would probably let the forest go ahead and grow into an upland hardwood hammock once it gets as far as this one. We actually have a few of these types around; they are sometimes beautiful examples of ecological succession, and we value them.

Here I am taking a look at a red-cockaded woodpecker cavity on Greenwood Plantation. Under many timber management systems, an older tree such as this one would not be allowed to remain in the forest, because it has ceased adding timber volume. But while mature trees might be inefficient from a purely economic standpoint, they provide critical wildlife habitat. Red-cockaded woodpeckers (RCWs) build their nest cavities in trees that are mature enough to have some red heart fungus, which makes the arduous job of excavating the cavity much easier for them. But they also need living trees, because the running sap around the cavity holes protects the birds from some predators. And when RCWs abandon these cavities, lots of other creatures will move in and use them. Scientists have found dozens of species making use of RCW nest cavities. Throughout the 1960s we estimated that the Red Hills plantations harbored about five hundred colonies of RCWs. We based this figure on an intimate knowledge of the land we worked, anywhere from two to three hundred thousand acres over the years. Unfortunately, new landowners were more interested in the monetary value of these beautiful and productive forests, and many of the old trees and the woodpeckers are gone.

I love being in the woods with Julie, and it's even more special when we are in the Big Woods of Greenwood, as we are in this photograph. Julie is an accomplished naturalist in her own right and an authority on butterflies. Her butterfly garden at our house has become a destination for butterfly enthusiasts, and she has consulted with many gardeners throughout the Southeast on designing butterfly gardens. Julie's approach to gardening is philosophically the same as the Stoddard-Neel Approach to forestry, and her knowledge and appreciation of nature have been an important influence on me and my practice. At the heart of our management philosophy is aesthetic beauty. We let natural processes shape our aesthetic sense of what the natural world should look like, and our sense of aesthetic beauty, in turn, acts as our guide to everything we do in nature. (Photo © Beth Maynor Young / Kingfisher Editions)

history—and then the visitors, who were either passing through or engaged in something around the territory, would be there too. It was a regular Sunday morning event. In the summertime we would sit outside, usually, and in the wintertime we would be inside, but the conversations were always animated. Along with discussions of the tower study and Mr. Beadel's search for something meaningful to do with his land, these regular gatherings contributed to a growing sense that there was an important scientific community in the Red Hills, one to which Tall Timbers would give an institutional home.

Mr. Stoddard had the first bird window—a large, wall-sized window that looked out on designed bird habitat with feeders of various sorts—that I ever saw, and it became a minor trend among the folks who were out there back then. He had it in place when Julie and I moved out to Sherwood, and we just fell in love with it. It certainly gives you joy to watch the birds and other animals that visit, but you also get a good sense of what is going on seasonally; it is both a literal and a figurative window on the natural world. Mr. Stoddard's window was a big attraction for visitors. Arthur Allen and Paul Kellogg, the founders of the renowned Cornell Lab of Ornithology, came down frequently to visit Mr. Stoddard and photograph at the window.

Following Mr. Stoddard's example, Betty and Ed built a bird window of their own. I remember when Betty built her first bird feeder, which was just a tray outside her kitchen window. She was raising two children, and basically taking care of Ed and Roy, so she was spending an awful lot of her time in the kitchen. And so she put a tray feeder right outside her kitchen window so she could watch the birds. Betty was a good naturalist, too. She was actually a trained botanist; she studied under Dr. Herman Kurtz at Florida State. Ed and Betty made a good team in that respect. A lot of Ed's success was due primarily to Betty. She kept him on an even keel, and stabilized him, so that he could keep his mind focused. Ed and Betty liked looking at the birds out that kitchen window so much that, later on, they created their own bird window in their old dining room. Following suit, Julie and I added a room to our little cabin on Sherwood that contained a little bird window. Then, in the design of our present house we included a large bird window in our bedroom. The bird windows were always a big attraction during these gatherings.

Famous ornithologists and naturalists gravitated to the "Sunday Morning Bunch" gatherings. James Tanner, who conducted the study of the ivory-billed woodpeckers in Louisiana's Singer tract, came by for a visit, and that's where I met him. He was visiting Mr. Stoddard and the Komareks. A lot of artists came by as well. Athos Menaboni, the well-known wildlife artist, and his wife, Sara, were good friends of Mr. Stoddard and got to be better friends of the Komareks. George Miksch Sutton, who was an important ornithologist and artist of the time, came down for multiple visits. Sutton illustrated Thomas Burleigh's *Birds of Georgia*, and he lived on Sherwood while he was painting those pictures. He also wrote the foreword for Mr. Stoddard's autobiography. Walter Stevens, a plant explorer and grass researcher over at the Tifton Experiment Station, came by as well. His nickname was Cowboy. He had been all over the world as a plant explorer, and he was a superb naturalist. Lucien Harris and his wife, Lucille, came down most winters, and they were always around, too. Lucien was a lepidopterist—the author of *Butterflies of Georgia*—and he really got Julie and me interested in butterflies. Julie is now a master butterfly gardener. Wallace Grange and his family also came to visit many times. Wallace took over Mr. Stoddard's job in the Biological Survey for a few years before he set up a game preserve and game propagating business in Wisconsin. He was a fine field biologist and naturalist. So the area around Sherwood and Birdsong was really a Mecca for naturalists, and you knew where to find them on Sunday mornings. These Birdsong gatherings involved a cross pollination between mammalogists, lepidopterists, ornithologists, ecologists, and others sitting down together and constructing a composite picture of how nature worked. It was fun, but there was a lot of serious discussion about what was going on in the field of natural history. These discussions were also the platform for the Tall Timbers idea.

After a long process, Mr. Beadel donated the land and a small endowment for the creation of Tall Timbers Research Station in 1958. But there was definitely a need to raise more money. Ed Komarek's connection with Greenwood Plantation, owned by the Whitney family, was important in regard to the funding of Tall Timbers. By the late 1950s Ed had become the manager of all the Greenwood properties, a total of four of them. Ed had been instrumental in developing a hybrid seed corn business at Greenwood, with cooperation from the Coastal Plain Experiment Station in Tifton. Mr. Whitney wanted to do something at Greenwood, if possible, to help southern agriculture. The Whitney family had enjoyed Greenwood for so many years, they had been good nonresident landowners, and they had been generous to the area. But they still wanted to do something more. Their goal was to develop and market hybrid strains of corn that fit the environmental conditions of the Southeast and would raise the yields of local farmers, as similar hybrid strains were doing in the Midwest. The Midwest was the nation's premier corn producer, and the South was falling behind.

Corn had long been an important crop in the South, but the average corn yield in the region was not great, in part because of how farmers got their seed corn. My father was a farmer, and by spring there would only be a small pile of corn left in our crib. It was ear corn, still in the shuck. My father would go through and pick out the best ears, which could not be deformed in any way, and that was the seed they planted for that year. He and the region's other farmers did the same thing year after year. They did not have the advantage of the high-yielding hybrids developed for midwestern farmers. As a result, the average yield in the area was about eighteen bushels per acre or something like that, as compared with midwestern yields that approached one hundred bushels per acre in the decades after World War II.

One of the big drawbacks to growing corn in our area was that we normally had a spring drought that coincided with the time the corn was blooming, and those droughts cut the yield. So Ed got it in his mind to study how Indian groups in the Southwest had adjusted corn to their arid environments. He, Betty, and the children made several trips over the next few years out to various western Indian tribes, and Ed literally had to join the Hopi tribe. They would not just give him any corn at first. They would not even let him look at it when he first went out there. He had to gain their confidence by becoming one of them before they started giving him a little corn with which to experiment. Ed was fascinated by their corn agriculture. They planted each kernel about eighteen inches deep in the arid desert environment in order for it to get the moisture that was below ground. Ed thought that if he could cross that dry-land corn with our southern varieties, he might be on to something. He worked with a scientist over in Tifton, Dr. Wayne Freeman, to get that gene out of the Indian corn and put it into corn over here. That effort did not take off, but Greenwood was successful as a seed company for a while. "Dixie 18" became an important variety in the South, and Greenwood made a lot of money distributing it, at least for a while. Yields in the area also went up a great deal, which was a tremendous benefit to the southern farmer. But, like many corn varieties, Dixie 18 was susceptible to leaf diseases because of our high humidity, and, on top of that, the big companies moved in and took over the hybrid corn business in the South. Greenwood phased out their seed corn business in the late 1970s.

Ed had been a faithful and successful manager for Mr. Whitney, and Mr. Whitney funded a lot of projects above and beyond what a normal plantation owner would fund for a manager. The development of hybrid seed corn was a perfect example. And as he was with the corn project, Mr. Whitney was generous to Ed in the development of Tall Timbers as a research station. He funded many of the first functions at Tall Timbers, which usually involved a barbeque at Greenwood, and he covered travel for Ed and similar expenses. Like most of us in those early years, Ed was also doing a lot of work for Tall Timbers that was on a volunteer basis, and the expenses came out of his own pocket. Back then, if you got a group of people together who had an idea, wanted to make it work, and volunteered wholeheartedly, you could get a lot done without much money. But Mr. Whitney was a tremendous help in providing us with some early operating funds.

Ed was also becoming a more visible personality in the community, which helped with spreading the word about Tall Timbers. He had developed a noontime television show for WCTV (whose tower was on Tall Timbers land) in the 1960s dealing with agricultural and land management issues. He went by the handle "Country Low," which derived from the idea that temperatures were lower out in the country. He made himself into an amateur meteorologist, and he would always lower the official temperature by five to ten degrees for the benefit of farmers. He had all sorts of people on that show, mostly farming authorities. And he even demonstrated how different types of fuels burned, on live television inside the studio!

It is hard to imagine today, in this era of big money science, that a research station could develop the way Tall Timbers did. Tall Timbers got going while we all had full-time jobs. Mr. Stoddard still had his timber management business, and he also was working on the tower study. Ed and Roy both had full-time jobs over at Greenwood. I was working with Mr. Stoddard in the timber work, and on other projects. As far as the early days of Tall Timbers, I was there to do whatever they needed of me. What little research I conducted for Tall Timbers was practical, done on a day-to-day basis, with only the land to show the results. The point is that all of the official members of the "staff" were also engaged in daily work, and while that sometimes got in the way, Tall Timbers also was a beneficiary of it, by design.

One of the first things we had to do after founding Tall Timbers was to create a board of directors. Mr. Beadel's lawyer was Will Oven, a fine man from an old-line Tallahassee family. Will was instrumental in drawing up the charter. He handled all of Mr. Beadel's legal work, he was enthusiastic about Tall Timbers, and he served on the board until he retired. We wanted naturalist professionals on the board as well. We got Dr. Malvina Trussell, who was a botanist at Florida State University, and Sam Grimes, who was an excellent amateur ornithologist and photographer from Jacksonville. Russ Mason, of the Audubon Society, and Roy Komarek were also on the board.

Tall Timbers differed in important ways from the second Quail Association. That association was really funded and driven by plantation owners. Tall Timbers, on the other hand, was driven by scientists and the questions we wanted to answer. Of course, the landowners were there. It could not have happened without Mr. Beadel, or the Whitneys for that matter. But their role was indirect. Nonetheless, there was a strong symbiosis between the practical land management we were doing on the plantations and the scientific work we were doing at Tall Timbers. We could not have done the latter without the former. Whatever we were doing with timber management on the plantations was not under the name of Tall Timbers, but Tall Timbers certainly benefited from the knowledge we developed in the field. We were getting paid for the timber work we did on the plantations, but they were not paying us for anything else. The benefit that Tall Timbers received from the fieldwork of Mr. Stoddard, the Komareks, and me, for instance, was the accumulation of knowledge about what we would today call ecosystem management. Every day was an experimental day. That is how ecological knowledge developed in the Red Hills, as the accumulated wisdom of lots of land management decisions and observations in combination with some focused scientific experimentation. And that is how I have continued to practice my land management. There is no universal formula for success in the art of ecological land management, only careful experimentation, adjustment, and what you come to know as beautiful and right.

Mr. Stoddard and I did not necessarily keep a lot of records or document all of our practical experiments on plantation lands, but we had the properties as proof of what did and did not work. When I marked a tract of timber on one place, and accomplished something better than I had been able to do before, I had that piece of land right there as my documentation. Of course, the problem with that is that much of what we accomplished on those lands is gone now. The land may or may not be owned by somebody else, but the trees have been cut and whatever we did there is no longer visible. One important exception is Greenwood Plantation, which was, unofficially, our main laboratory. The Whitneys were practicing good quail management there, as well as good land management in general. And Greenwood was the most successful quail plantation year in and year out. Ed had records that proved this. The land at Greenwood produced the most quail and the most timber of any plantation down here. They had several different properties with lots of diversity—over seventeen thousand acres total—so they could do a wide variety of things with it. There might have been some other places we were working on that equaled it, but year in and year out Greenwood was the premier property. Many people now say that you cannot have timber and quail on the same land, that you have to cut most of your timber to have quail, but Greenwood is our proof to the contrary.

Mr. Stoddard had a terrific and productive working relationship with the Komareks, which served Tall Timbers well. Both Mr. Stoddard and Ed Komarek liked the notion of a private research institution; neither particularly relished the prospect of fighting a bureaucracy, which is what would have been involved if they tried to ally themselves with another organization or state entity. If Mr. Stoddard wanted to do something, he wanted to just do it. He did not want to go through the procedure of requesting permission or getting funding. If that meant he did not have a lot of funding, so be it. If he wanted to get in the car and go somewhere, he paid for it; he did not care whether anybody paid him mileage or if he got permission to go, or if it was official or unofficial. I think the main thing with Ed and Mr. Stoddard is that they thought they could get more accomplished working as individuals. They were field naturalists first and foremost, and they did not need a lot of money. As long as they controlled Tall Timbers, they never had a lot of money, but they got a lot done. If they had been wealthy, there is no telling what they could have accomplished.

In the early years Tall Timbers had only a skeletal staff—it was not a big operation. The first person that we hired was a bookkeeper, a young man that Ed knew who had worked in the office at Greenwood. His name was Tom Quick, and had grown up on the Heard's Pond property, which the Whitneys had bought around World War II. Tom's father looked after the Heard Place for the Whitneys. Mr. Beadel had a management staff on the plantation, and they assisted us with the early research, but they were already on his payroll, so we did not need to pay them. Pless Strickland was the land manager, and Leroy Collins was his assistant. Then Mr. Strickland died, and Leroy managed there for a while until he died. Ed was the secretary-treasurer of the organization, but he received no compensation. Roy was on the board and always around to do what needed doing. I was the forester. Mr. Beadel was president of the board, and Mr. Stoddard was vice president. All of us worked for nothing, so there was no expense as far as that goes. We worked out of Mr. Beadel's house for a while. Then Richard Tift, Mr. Stoddard's business partner in Albany and a good friend, got General Richard King Mellon, who owned Pineland Plantation, to donate some money for an administrative building. We built a small building, called, appropriately, the Mellon building, which houses the Tall Timbers archives today. That was the first building we had, and everybody moved in. Things grew from there.

During the early years of Tall Timbers, in the late 1950s, I was deeply involved in maintaining and improving the timber business that Mr. Stoddard had started. I did it for selfish reasons—to keep myself employed—but I also did it to keep the land we managed in good shape. He had worked so hard on maintaining these places, and I felt a huge responsibility to keep them up. So that kept me busy. I was available at Tall Timbers, but most of the time I was tending to our forestry business.

Mr. Stoddard, who was in his late sixties at this point, was moving out of the timber business and devoting more time to his scientific research. The last stand of timber he marked was on Mill Pond Plantation in 1956, and it happened to coincide with a major shift in the lumber market. We marked and sold a lot of low-grade, number 3 timber that made boards in those days. Boards were one-inch planks used for subroofing and subflooring; they use plywood today, of course. The board market made a tremendous difference in our plantation timber management. With a healthy board market, we were able to sell the low-grade timber, and we received pretty good money for it. Then the market broke. Just after we marked that tract on Mill Pond, the market just collapsed. Plywood was beginning to replace boards, and it was a case of product oversupply.

The plummeting board market hurt us for a couple of years. The bids came in on the Mill Pond tract, and they were so low compared to our evaluation that we no-saled it. We just withdrew the timber from the market, and that tract just sat there with marks on the trees. That was something we rarely did. Because of his age and his work on the tower study, that was the last tract of timber that Mr. Stoddard ever marked. He ended up, over the course of the next year, turning the whole thing over to me. He still was involved, of course, but I did all the work.

My response to the low bids was to offer the Mill Pond tract for another sale the next year. When we sold timber, we were dealing with local people, and there were several good mills around at that time. But they had gotten to the point where, together, they could tell what we were going to do and what the market was going to do. It was, in effect, a kind of accidental collusion, which worked to keep prices fairly low. We were at a pretty distinct disadvantage until I made contact with Victor Beadles, of Beadles Lumber Company in Moultrie, Georgia. Victor was a young fellow just out of school at the time, and he did not know much about the business. In other words, he was not indoctrinated to their ways. Victor is a very good friend of mine, and a good conservationist too, and I will forever admire him for what he did for us.

I had already sold him some timber from the Albany region, so I got Victor to come in and bid on the Mill Pond tract. And we involved Arthur Tyson, a logger from Albany, who was a fine man. He was one of the old-fashioned loggers, and although I expect he cut some corners in his day, he did not cut any with us. We became good friends as well, so I arranged with Mr. Arthur to log this timber for Victor. We worked the deal out that way and then got Victor to bid on it. And Victor was the high bidder, because the other bidders submitted their same bid from the year before. It was sort of a standoff with them. They thought they had me, and they probably did in the long run, but they did not get me that time. Anyway, that was the last tract of timber Mr. Stoddard marked. From that point on he spent most of his time involved in the tower study and other things at Tall Timbers.

While the board market was sorting itself out, I went into the tree-planting business for a few years. We did not really sell any timber for about two years, so I had to make a living somehow. I bought a tree planter with Julie's brother, Jimmy Greene, and we planted millions of trees all over the state. Jimmy lived in Americus, Georgia, so he had contacts up there, and I had lots of contacts around Thomasville and Albany. We were fortunate that the board market broke during the years of the Soil Bank Act, which was a federal program that encouraged farmers to take land out of agricultural production and put it into pine trees. So there was plenty of business for anybody looking to plant trees. We mostly planted slash pine—"Soil Bank Slash" as they called it. I wish we could have planted longleaf, but there were no seedlings available. The state nurseries were only interested in developing the so-called fast-growing species like slash and loblolly. Anyway, there was a lot going on while we were trying to get Tall Timbers up and running. It's a wonder it happened at all.

In the early years, Tall Timbers was a personality-driven institution. That was one characteristic that made it work. You had a small and committed group of people who were experienced in working with and managing this landscape. They were working toward the goal of having a fine private ecological research station, and they were willing to sacrifice to do it. That lent the whole thing an air of informality. It is important to remember that none of the major players had formal scientific credentials. Mr. Stoddard did not even have his high school diploma—although he had become a published and respected field scientist—and neither of the Komareks had even finished their undergraduate college degrees—although Ed later received an honorary doctorate from Florida State University. So setting up a private institution probably made sense in that way as well. But they also recognized that they needed to create structures that would allow the institution to mature and be perpetuated beyond them.

Their most serious interest early on was in fire ecology, and that became a focus of on-site research. Mr. Stoddard had proposed setting up a fire ecology research station much earlier in his career, but he got little response from public or private interests. But with Tall Timbers, Mr. Stoddard, and especially Ed, ran with the idea to make it a national and international center of fire research. They realized Tall Timbers could make a mark by focusing on fire ecology.

The fire plot work was the first truly scientific research directly sponsored by Tall Timbers. Jimmy Greene and I, under Mr. Stoddard's direction and supervision, installed a series of plots at Tall Timbers that were then subjected to various types and frequencies of fire. We surveyed them, mapping every tree using a plane table, which is a field tool for surveyors to draw accurate maps. The plots were designed to serve as a demonstration of what happens under certain established fire regimes. There were different plots with varied burning regimes—summer burns and winter burns; one, two-, three-, and four-year fire cycles; and so forth. The plots were across diverse landscape gradients so that we could determine what happened in particular habitats. Mr. Stoddard spent a lot of time locating these plots so as to get the best results. We went out there and measured everything. For instance, we measured every tree within a three-quarter-acre plot and followed that plot over a long period of time to see what happened to the trees and other vegetation. You cannot always get a true demonstration of fire's ecological effects on a three-quarter-acre plot, but you get some sense. We knew what was going to happen to these plots for the most part, based upon lots of experience. But these experiments proved the point for those with less experience, and they gave our intuitive knowledge experimental form. It was a controlled experiment, but it also functioned as a visual demonstration of the value of fire and what might be achieved under different fire management approaches.

Beyond such basic research, the first thing Tall Timbers did to make a name for itself as a scientific institution was to start the fire ecology conference series. The first conference was held at Florida State University in 1962. Ed did the groundwork, and Florida State agreed to host it. They put their stamp of approval on it and gave it some academic legitimacy. But they knew that nobody was going to publish the results, so Tall Timbers had to publish them. Actually, Ed and Mr. Stoddard wanted to publish them out of Tall Timbers so we could retain editorial control. They had had enough of being edited when it came to fire. There was not much of what we would call science in any of the first few conferences, but they brought together a tremendous amount of knowledge and experience in land management using fire, and they had the surrounding plantation landscape to reference as proof of some of the things that Tall Timbers was recommending. As it turns out, these early conferences, and their published proceedings, were critical to encouraging and documenting major attitudinal changes about fire and fire management.

I was at the first fire ecology conference, and I was impressed, of course. I was impressionable too. I was thirty-five years old by that time, which is not necessarily young, but it was something to be surrounded by such luminaries. I was just a fly on the wall at that first meeting. So much of the larger policy discussion was brand new to me, even though I had listened to and been involved in the conversations between Mr. Stoddard, Ed, Roy, and the others that were involved. I do remember that Mr. Stoddard was a little anxious before the conference. He had received some fairly rough treatment regarding the use of fire, not only from the Forest Service but from academicians here and there who had taken the opposing view, and he was sensitive to the criticism that might come from others in the forestry and land management profession. Fire was still a no-no in the late 1940s at the University of Georgia when I was there, for example. We had a fire course, but it was on how to extinguish fires in the woods. I imagine the same was true at most forestry schools in the decade or so between my graduation and that first conference. So even though Mr. Stoddard had fought tough battles over fire's ecological role in the 1930s, there was still plenty of resistance out there. Ed Komarek was sensitive, too, but not to the extent that Mr. Stoddard was. They went into this fully expecting a great deal of criticism, but they did not receive much. I think they handled it well.

The first conference was designed to re-create, as much as possible, the history of fire ecology from the perspective of the older people still living. Roland Harper was one of the speakers, and I will never forget meeting him. He was a famous botanist in the South who had argued early on that the longleaf-grassland system was fire maintained. Roland was a bit of a renaissance man. He was born in Maine but came to the South early on when his father landed a teaching job in Americus, Georgia. His brother, Francis, was also a famous botanist who brought the Okefenokee Swamp to everybody's attention. Roland graduated from the University of Georgia in 1897 with an engineering degree, but he remade himself into a botanist in the next few years; he received a PhD in botany from Columbia in 1905. By the time I met him, he had worked with several southern states in various capacities—as a botanist and geographer in the geological surveys of Georgia, Alabama, and Florida; as the head of the Florida state census; and as a professor of economics at the University of Georgia. He also kept an office at the University of Alabama for many years.

I had heard about Roland from Mr. Stoddard all the years that I had been with him. Roland was one of Mr. Stoddard's favorite people, because he was a doer. He would travel to some place like Chattahoochee, Florida, by train, according to Mr. Stoddard, and he might walk miles to a community where there was a report of some kind of plant that he wanted to see. He did not have a car, and he never learned to drive. If he could catch a ride on a mule and wagon he would, but he did a lot of walking. And I think that is a big reason he became such a stellar naturalist. He would walk to where he was going so he could see nature from the ground level. People do not walk like he did anymore. Roland gave a pretty good history of the use of fire at the first conference, and he looked just like every picture I have ever seen of him. He had on a black suit with a high white collar and string tie. That suit is all I ever saw him in.

There was a broad audience at that first conference. They were not just those who had always supported fire. One of the presentations was by R. A. Bonninghausen, who worked for the Florida Forest Service. He came after Stoddard, Beadel, and Harper, who had all given historical papers that were pro-fire. Early in his career Mr. Bonninghausen had apparently been a fire-excluder, and in the first paragraph of his talk he had the tone of "well, I'm not real sure what I'm doing here." There was a little conflict between the two sides of the fire divide, but it was a good meeting and served a constructive purpose. The fire ecology conferences are probably the greatest thing that Tall Timbers has done as a research station, and the early 1960s was a ripe moment for putting these voices together.

From the beginning, Mr. Stoddard, Ed, and others had a sense that this first fire ecology meeting would be a springboard for pushing fire ecology out into other landscapes, other parts of the United States, and even other parts of the world. Their experience was particular to the southern coastal plain, but they had a broader vision. Mr. Stoddard was most interested in establishing fire as a legitimate ecological tool in the longleaf belt of the Southeast, but Ed had a vision of promoting fire worldwide. And he eventually did. We had a conference that addressed fire in Africa, and we actually held conferences in California, Texas, Montana, Oregon, and New Brunswick, Canada, sometimes in conjunction with other organizations. A lot of good information was shared, and we changed a lot of people's minds. So the work on fire in the southeastern coastal plain was a major catalyst for a broader ecological reconsideration of fire, both nationally and internationally.

Ed Komarek deserves a lot of credit for expanding the research program at Tall Timbers and promoting fire all over the world. He and Betty started making international contacts with others who were interested in fire management, and they traveled extensively to see other fire regimes, especially to Africa and the Mediterranean countries. We have had Portuguese foresters over here. They burn their Mediterranean pine, which, like our coastal slash pine around the Gulf coast, used to rim pretty much the entire northern Mediterranean, I think, but today there is not much left. The African fire conference was very interesting, and we made good friends with some scientists from Rhodesia, before it became Zimbabwe. The African scientists were far ahead of their time, as far as we were concerned. They were ahead of us in a lot of respects, with their use of fire.

The Tall Timbers Fire Ecology Conferences had a strong influence on the management of lands in the American West as well. This is jumping ahead a little bit, but in 1972 there was a task force that went to the Apache reservation in Arizona to study the effects of fire in the ponderosa pine. Harold Biswell, a friend of Mr. Stoddard and, eventually, a major proponent of controlled fire in the West, led the task force. Dr. Biswell was studying grasses over at what is now the Coastal Plain Experiment Station in Tifton, Georgia, and that is where he learned about fire. He then went to California, to teach at Berkeley, where he began to argue for the need for fire in the West's Ponderosa forests. He wrote a booklet on the subject, *Ponderosa Fire Management*, which Tall Timbers published in 1973, and he later went on to publish an important book on prescribed burning and its ecological importance to California land management.[3] Incidentally, the man who occupies Biswell's position at Berkeley today, Dr. Scott Stephens, brings a crew of students over here occasionally. He sent me the booklet that Biswell wrote about fire in the ponderosa, and he told me that Biswell was never fully accepted at Berkeley because of his fire beliefs. I had met Dr. Biswell at the 1967 Tall Timbers Fire Ecology Conference in California, and he was certainly a brilliant scientist. We watched him put on a controlled fire demonstration in the California mountains, and that fire stopped right at the top of a ridge, just like he wanted. He was an expert at using natural conditions to control fire.

One good example of the differences between research at Tall Timbers and the fire research done elsewhere is in the comparison with the work done by the Forest Fire Laboratory in Macon, Georgia. The historian Stephen Pyne made this comparison in his book *Fire in America*, and I think he made a good point.[4] Mr. Stoddard always referred to his work, and by association Tall Timbers' work, as applied ecology, and everything I did was applied. Ed established a close relationship with the Fire Lab in Macon, and there was some meaningful interaction between the two institutions. They attended a lot of the fire conferences and made presentations. But they were studying the physics of fire—the actual temperature of the flame one foot above the ground, for example, which is your killing area when you are burning pine trees. They learned all that, and that was good research. However, we learned it intuitively the first time we killed a tree out there. We did not necessarily need to know the exact temperature to know what it took to kill a tree. When it came to fire, at Tall Timbers we were more interested in applied ecological science than we were in determining the physics of fire.

The 1960s were also important years for the rise of the environmental movement, and while Tall Timbers was by no means a focal point of that movement, the institution made a contribution. For instance, Rachel Carson—whose *Silent Spring* (1962) laid the groundwork for the environmental movement—was discussed many times. Mr. Stoddard had long been involved in fire ant research, and he adamantly opposed the USDA's broadcast spraying campaign aimed at fire ant eradication—a campaign that Carson skewered in her book. So we took notice when her book appeared. Mr. Stoddard was also connected to some important figures in the fields of conservation biology and ecology, disciplines that were gaining prominence as Americans became more environmentally aware. He knew Archie and Marjory Carr, for instance. Archie was a Floridian who did important sea turtle research and was one of the founding figures of the field of conservation biology during this period. Marjory was also an accomplished ecologist and ornithologist. And, of course, Mr. Stoddard knew Eugene Odum quite well during a time when Odum was gaining public recognition as a key architect of the field of ecosystem ecology. The point is that, as Tall Timbers matured, it did so in the context of a larger sea change in American environmental values. The conversation we started about fire fed into that.

The Endangered Species Act (ESA) of 1973 was also an important piece of legislation in relation to the longleaf region and the work we were doing at Tall Timbers, because it highlighted the need to protect keystone species and their habitat. The most important example of this has been the red-cockaded woodpecker, which required longleaf and other pine forests with older trees. By the 1960s and 1970s most of the remaining isolated red-cockaded woodpeckers around the state were in little churchyards with a few old longleaf trees in them, or in stands of timber where some old trees had survived. Their presence was almost accidental. But the ESA provided those of us who had long been concerned about the total habitat of these forests with some leverage and an additional argument for protection. And, as I explain more fully in the next chapter, the Stoddard-Neel Approach is well suited to producing and perpetuating forests where red-cockaded woodpeckers and other endangered species can do well. The ESA has had a few negative effects, as some owners of good longleaf lands have cut their mature timber because they feared that red-cockaded woodpeckers would take up on their lands and thus limit their freedom to manage their properties as they chose. That's a problem, and we need to work on that. But for the most part the ESA has helped to highlight the importance of the longleaf biome and the distinctive species it harbors.

But in other ways our work swam against the preservationist tide of the era. The late 1950s and early 1960s was a period of growing support for the preservation of wilderness, and Americans of all stripes were increasingly committed to preserving pristine landscapes from human exploitation and manipulation. That effort peaked with the passage of the Wilderness Act of 1964. It is important, in the context of such preservation efforts of the time, to understand that the goal of Tall Timbers was active habitat management. We were not merely preserving a static landscape untouched by past or even present human activity. We were actively maintaining a system that would have degraded without our interventions. As Ed once said, "Another major resolution in the founding of Tall Timbers was to place great emphasis on habitat management so as to have better nature management in place of the outmoded 'nature conservation,' which has become virtually ineffective and meaningless."[5] We were all for preserving wilderness, but in this country down here, the best way to preserve something is to get out there and actively manage it. Ours was a form of ecological experimentation and manipulation; we manipulated the environment with the goal of preserving native biodiversity, often within a working landscape.

I think one of the different things about Tall Timbers and one of the keys to its success was that the laboratory was the land itself. We were able to do small, simple things on the lands we managed that made a big difference in terms of wildlife. Let me give you a case in point. Ed had a piece of cutover woodland on Birdsong that he converted into a pasture. He cleared the scrub hardwoods and a few pines to create an open pasture with just two or three trees per acre. But in doing so he was careful to leave several upland black gum trees for the fruit that they produced. The fruit is excellent birdfeed for many species of birds, including wild turkeys and quail. Mr. Stoddard almost certainly had pointed that out to him, because one of the first things that I learned from him was that if you had an upland black gum (*Nyssa sylvatica*) in open pinewoods, especially a seed producer, you wanted to keep it for the birds, even if it served no economic purpose. We would actually cut pine trees from around it to keep the fuel load down so the fires would not kill the black gum. It was a bit of forest diversity that you wanted not just to keep, but to cultivate. I have often seen red-cockaded woodpeckers feeding on black gum fruit from trees that we left in the woods, and, of course, quail eat the fruit if it falls to the ground. That pasture was one of Mr. Stoddard's favorite birding places after Ed got it in shape and all of these things were producing. That is where he went to find migrating orioles, because he rarely found them out in the woods. You might luck upon one somewhere, but during migration the best, most predictable place to find orioles was in that pasture with the black gum fruit.

That is an example of the habitat management to which Ed was referring. I call it management, but I guess you could call it manipulation. To do it right, first of all, you need to have the knowledge of what you are doing and what your goals are. You have to know which tree is preferable to a certain species. Then you have to know what you are doing as far as removing fuel from the base of black gum trees on land that you are going to burn. For such management to be successful, you have to know a lot to begin with, and then you have to keep observing to learn more. In that way, such working knowledge can become scientific knowledge, though it can never be completely abstracted and separated from the specifics of the landscape.

I was taught by an expert, who saved me a lot of time, and I would not have figured out all that I know today without such expert guidance. But because I was taught this, and then because I have put it into practice and seen it work, I can do it. I can recommend such habitat management techniques to a client who may or may not have much interest in migratory songbirds. I can show them that they can help the environment if they do just a few things. You might be able, through experimentation, to develop the science to where you can say that you need six producing black gum trees per twenty acres, or something like that, to increase certain bird populations without impinging on effective fire management or other economic ends. You can get to a point where you can create a formula, in other words. But if you are paying attention, do you really need such a formula? And even if you have a formula, not many people are going to follow through unless they have a personal interest in it.

There was little indoor laboratory work going on in the early years at Tall Timbers, except by people such as Bruce Means. Bruce was a graduate student at Florida State who made contact early on to do research with Tall Timbers, and he became one of our earliest doctoral fellows. He eventually became the director in 1978. Bruce was studying rattlesnakes at Tall Timbers, of which there were quite a few. He studied the snakes in the field and in the laboratory. But the snakes occurred out on the land, so even his research was only partly laboratory based. He put radio transmitters in many of the rattlers he captured, and, after releasing them, he was able to track their movements over a large area. Such research was just a natural outgrowth of the work of people who wanted to learn more about the land and how to best take care of it.

In a sense, what we were trying to do in the early years of Tall Timbers was not merely to put in place a scientific management program that protected the existing lands of high ecological quality in the Red Hills, but also to pursue a science of perpetuation and restoration. Environmental perpetuation is a little different from simple preservation, because environments are always changing. From a practical standpoint, the great majority of landowners were not even aware of the ecological restoration we were doing unless Mr. Stoddard or I told them what was going on and what we were doing on their land. But often, when we managed plantations, we explicitly sought to restore certain habitats.

Restoration included burning brush land, and it sometimes required severe fires, particularly in places that had not burned in a while. The whole process necessarily had to go on and on; you never can restore something and then dust your hands off and say I am through. You better come back next year and keep restoring it. And that speaks to an important misconception that many people—including some scientists and land managers—have about restoration. To restore something seems to imply some sort of fixed state toward which your management actions are aimed. But what we were after was some combination of ecological function and economic or aesthetic utility. In that sense, restoration was not only a constant management task, but a constant education as well.

Early on, Ed Komarek said of Tall Timbers that its staff "would not strictly engage in specialized study projects but would, in the spirit of man's greatest scientists, devote their lives to the study of nature as a whole. . . . [T]he emphasis was on a search for knowledge, and an understanding of nature as ends in themselves."[6] Ed went out of his way to deemphasize specialization. He and Mr. Stoddard rebelled against organized, textbook management. You have to make decisions on the ground if you are a good manager. You might have to change your burning regime because of certain conditions created by weather over time, for instance. That is why it is hard for me to draw a specific management plan for a property. I cannot draw one. I can give you a general idea of what I am going to do on a given piece of land. I am going to burn this property every year. That is my management plan. Now, that means sometimes I burn 20 percent of it, or sometimes I will get lucky and burn 95 percent of it, but either way I will have to adjust what I do the following year based on those results. It is not a scientific way of looking at it, but it has worked for me, which makes Ed's take on conservation and management all the more pertinent. Anytime you become stuck in a rigid management plan you potentially put yourself in a straightjacket; the land can change on you quickly, and you have to be ready to adjust.

Ed also wrote, "An understanding of natural principles and processes does not necessarily require formalized scientific processes. . . . Science is not solely an accomplishment of modern man; it developed as a continuum from earlier times."[7] I agree with that 100 percent. I am not a scientist, but I have been around a lot of scientists, and I am close friends with many scientists for whom I have the greatest admiration. Yet, if you put them in charge of managing Greenwood or one of these other plantations that is still in good shape, most of them could not do it. They might call themselves land managers, but over a short period of time—just a few years—you could begin to see differences in the forest. Or else it would cost them an awful lot of money to avoid that. It is very difficult to be an academic scientist with a great reputation and still be a good land manager. They are very different jobs with different skills. I think it takes as much time to do one as it does the other, and while both are important, we should not presume that good academic science can be a shortcut, a replacement for good land managers steeped in experience. At Tall Timbers we came to our science through our land management, and they reinforced each other.

Some of the early participants at the Tall Timbers fire conferences were concerned about whether the institutionalization of our form of management might infringe on property rights. We have always had to be aware of property rights down here, and we were never interested in forcing landowners to do something they did not want to do. That said, I believe regulatory oversight is sometimes necessary. I know this: if you take every owner of wild land—and by wild land I mean any land that supports wildlife reasonably well, whether it is a farm, a ranch, a forest, or whatever—in the United States of America, I am not sure how many would be good conservationists. So sometimes it is necessary to try to find a regulation you can apply to keep people from destroying their land, while still not infringing too much on their property rights. Forestry is a good example. We have proven that you can have a legitimate, reasonable harvest of timber off of land and still maintain most of the other natural features that the land is capable of producing. Sometimes you need to use extra care—like with the red-cockaded woodpecker—and in those cases, regulation might be the best way to manage the situation. There's a certain ethical responsibility that landowners have when they have a red-cockaded woodpecker population on their land, but many are not going to care. I do not much like government intervention either, but I think it is necessary in those cases. One of the goals of our management system is to keep the land in such good shape under local control that you are ahead of the regulatory curve.

To a certain extent, then, we had to help landowners understand that the goal of Tall Timbers was not to use science to impose a set of restrictions on their land use. But we also thought we could convince them that, in terms of timber and quail management, they could have the best of both worlds. They could have quail in good numbers, harvestable timber and accruing timber value, and land that contained good plant and animal diversity. In most cases in dealing with my clients—and I was an unofficial representative of Tall Timbers even when I was practicing my business—I could recognize any potential problems that might arise between ecological preservation and management goals, and often you could at least balance the two. Sometimes if the clients were adamant, then I had to back off and let them maintain their position. Their position was, in most cases, one of private property rights. They were staunch private property rights people, and they wanted to be able to do what they wanted with their land.

Getting fire management principles disseminated to landowners beyond the quail plantation set was more of a challenge. Most of the local clients and supporters of Tall Timbers had the impetus of bobwhite quail propagation as their selfish motive, and they had been taught for over thirty or forty years that fire was absolutely essential for keeping quail numbers relatively high. Even if you are releasing birds, you have got to burn your woods. But when you get out into the fringes of the plantation regions, it can be difficult to convince somebody with a thousand acres to implement a specific management program. Most people with a thousand acres are not multimillionaires, and careful ecological management takes time. But you can still go into a program where you can have trees and some quail, if you are wise, not greedy, and understand the management. You do not have to cut all your timber unless you are stone-cold broke, and even then, that is just a delay tactic, because once you cut all your timber, then what are you going to do? Unfortunately, what has followed too often from such a situation has been the conversion of longleaf lands to plantations of pine planted for pulp and paper. And there is little ecological diversity, or aesthetic quality, on those lands.

Harry Beadel died in 1963, which was a tragic loss. But, luckily, his passing did not have much of an impact on the operation of Tall Timbers. Mr. Beadel had already given most of the Tall Timbers property to the research station prior to his death, though he lived on the property full time in the last years of his life and kept some of it—his house and part of the property—for himself. At his death, the rest of the property transferred over, and what endowment he had left was there as well. Mr. Stoddard was still there and the board was there, so when Mr. Beadel died there was little chance that the philosophy at Tall Timbers would change. It was only after Mr. Stoddard and Ed passed that the board was gradually replaced and Tall Timbers began to change.

As I have intimated thus far, Ed Komarek was really the energy behind Tall Timbers during the early years, while Mr. Stoddard was the applied ecology and land management brains of the operation. Ed's brother Roy played a less public role, but he was nonetheless an important figure in the organization. Like Betty, Roy was a stabilizer for Ed. Roy was quiet. Ed could be egotistical from the standpoint of trying to acquire more knowledge all the time, and then promoting that knowledge for others to acquire. There is nothing wrong with that, because he was really trying to get his ideas out to the public. Roy did not ever care anything about giving a talk at a meeting or anything. He just went about his business, which was good land management. He did most of the land management at Greenwood while Ed was off building Tall Timbers. They made a good team in that respect. Ed was fortunate to have Roy, and, of course, Roy was fortunate to have Ed.

I do not remember any serious changes to Tall Timbers in terms of its goals and management approaches during the early years, but I do know this: as the fire conferences developed and were held annually, and as Tall Timbers grew in importance, more doors began to open. Tall Timbers developed a good working relationship with Florida State University. We made an effort to cooperate and induce cooperation from some of the biologists down there, and some of the entomologists as well. Walter Tschinkel, who is a well-known entomologist at FSU, did some work at Tall Timbers during the early years. He was always on call, unofficially, if we needed to have an insect identified or to get a life history of a certain species. As time went by, Tall Timbers began to develop deeper connections with the larger scientific community and to blossom with intellectual opportunities and connections.

One of the results of this broadening of interests was a much wider array of publications produced by Tall Timbers. Ed became interested in earthworms, for instance. He would find something that little was known about, and he knew enough to know that there was a lot to learn. That is what he would jump on. So when he became interested in worms, he went up to Harvard and met Dr. Gordon Enoch Gates, who was the foremost earthworm expert in the United States. He had an illustrious career and a large number of publications. When Ed found him, though, he was long since retired. He was only allowed to maintain a desk and his files, but they had moved his desk out into the hall. Dr. Gates, this elderly expert on earthworms, was being swept aside. Ed got to know him and befriended him, and Dr. Gates became an associate at Tall Timbers. He had a wealth of knowledge on earthworms, and Ed knew enough to tap his knowledge, both for his research and for the research station more broadly. Mr. Beadel had a similar idea in the early years of Tall Timbers, even probably before it was founded. He recognized that if a prominent scientist in a field that fit in with the focus of Tall Timbers was forced into retirement, they could make a place for him at Tall Timbers so he could continue his life's work without fear of restrictions on age.

Ed also set up what we called the Beadel Fellowship Program. Ross Arnett, a distinguished entomologist who had worked for the USDA and a number of universities, and Will Whitcomb from the University of Florida, another entomologist, were early Beadel Fellows and did some research at Tall Timbers. We became quite interested in entomology, because nobody knew much of anything about the insects of this system and how they fit in. Mr. Stoddard had done food studies in relation to the bobwhite quail, which had revealed some of the region's insect diversity, but there was so much more work to be done. Ross Arnett came to us from Purdue University, I believe, and stayed at Tall Timbers for three years as a Beadel Fellow. There was no formal way to keep him, and he went on down to Florida later.

One reason we became interested in entomology was because we had some practical management studies on disking old fields to produce certain vegetation at certain times of the year, which in turn attracted certain species of insects. We were looking for insects that would feed on crop pests, as an alternative to using chemical pesticides. We hosted several conferences on the subject, and those studies were ongoing up until a few years ago.

Today we recognize that one of the great features of a relatively undisturbed and well-managed piece of longleaf land is the tremendous diversity in its understory. By some measure it is among the most diverse natural systems on the planet. We all had some sense of that at the founding of Tall Timbers, but it was also an ongoing process of discovery that was at the center of our research agenda. When I first joined Mr. Stoddard, he really hammered home a simple lesson: the native ground cover, where most of that diversity resides, occurred primarily under longleaf. It fit into the natural longleaf ecosystem, which was easily reproduced continuously, restored even, if you used fire. There is an old misconception that the wiregrass understory is sterile, but if you go out to the wiregrass and look around, there is diversity everywhere. Old-field land was more conducive to carrying a larger population of bobwhite quail, because the old-field successional processes produced good habitat for them. That is not necessarily true of all old-field land; the soil is so depleted on some old-field land that quail production is a struggle. But over time, as we worked both with old-field land and land that had not been plowed, we began to recognize the importance of understory diversity and to lend it new value.

One of the things that prompted this was when the paper companies moved south and started buying land and destroying thousands and thousands of acres of longleaf land with native groundcover. They replaced these lands with monocultures of planted pine. Their process of site preparation was particularly destructive. They would broadcast-spray herbicides to kill all the vegetation, and then plant the pines in bedded rows. That process destroyed an unknown number of rare and desirable plant species, and it made the land unlivable for many animal species. That was shameful, and it required a defensive response.

The early years of Tall Timbers coincided with the period of peak production in my business, which was managing timberlands from an ecological standpoint. I had, of course, gotten into that groove by about 1960. As I noted, that was when Mr. Stoddard backed off of marking timber. I had to hold everything together on my own, which was not easy. I had someone working for me, but that was not the same as working with Mr. Stoddard.

Mr. Stoddard died in 1970, and that created a void in our lives. He was always a pleasure to be around. In the last year or so of his life, as you would expect, his memory was not as good as it used to be, but he still had a tremendous interest in everything, and he particularly enjoyed visiting Julie and sitting at the bird window. Of course, I felt increased pressure in his absence, even though I had been carrying on the business for several years.

After Mr. Stoddard died, it took us about a year to build and move into our new house, which was completed in 1972. Mr.

Stoddard had wanted us to stay at Sherwood indefinitely, but we did not feel comfortable doing that in his absence. But, kindly, his son, Sonny, allowed us to stay there until we got the house built on the Thomas County family farm. So we were busy. I was working just as hard as I could to carry on Mr. Stoddard's land management legacy. Marking timber is not easy, especially when you mark the way we mark. If all you have to do is make a decision about the number of trees and board feet you are going to take, then you can do that without thinking. But if you have to go out there and assess every tree individually in relation to all the things that we are worried about, and then make a choice as to which ones you can spare out of a group, then it is more difficult. So we were busy during that period.

Julie was raising our daughters, too. We were lucky to be able to raise them in the woods, and they both liked it. They used to burn with us when they were little. They loved to burn. Julie, the oldest, also loved to hunt. Susan enjoyed going along, but she never did like to pull the trigger. Julie killed her first deer at fourteen, and has killed several more since. She has not hunted in the last few years, but she loved to hunt. I took them turkey hunting many times, and they both shot a turkey. They loved to fish as well. They were and are great children, and we had a great time.

The Komareks continued working at Greenwood the whole time they were building up Tall Timbers. They were there several years after Mr. Whitney died in 1982, but Ed was in a terrible automobile accident in the mid-1980s. He was running around more and more before he had this accident, trying to catch up on some things, knowing that life was getting shorter and shorter. As you get older, you cannot help but think about all the things you need to do or want to do but have not yet done. He finally reached the point where he retired. Mr. Whitney died, and with Mrs. Whitney in charge of the place, they were not hunting the property like they used to. They retired Ed because he truly was not capable of being in that position anymore after the accident. Then just a short while later—a couple of years—they retired Roy, because they had quit using the property like they used to. They phased him out, and really, from that point on they did not have a real qualified manager. I have been looking after their timber and have been a sort of unofficial land management consultant. They have a fine crew that maintains the place, but Ed and Roy have proven difficult to replace.

As far as my personal involvement in Tall Timbers goes, I was always there and I participated, but I was in the background during the early years. I did not think I had the qualifications, or the ability, to become a leading figure in the organization. When Ed died, I did step forward, and Roy and I worked together as a team on the board of directors to perpetuate the values of Mr. Beadel, Mr. Stoddard, and Ed. Roy and I dominated the board, and we had a pretty good staff. We let the staff go the way they were going. We tried to keep up with it, and I recognized that it might not last for long, so we tried to get some people on the board that we thought would have the philosophy to carry the idea forward. But we got some third- and fourth-generation plantation people on the board who did not fully understand or appreciate the founding vision or mission of Tall Timbers and wanted to take it in a different direction. And as soon as the new board members got control, they got rid of me. That was in the early 1990s. It was very frustrating for me to leave Tall Timbers. We had worked so hard to build that place up, and I felt like I was ready to help move it forward. But I just could not see eye to eye with the majority, so rather than have a constant thorn in their side, they let me go. I had grown rather tired of being the thorn anyway. They began to shift the focus from ecological management to intensive quail management to the detriment of the total ecology of the plantation landscape in the long term, and I could not be a part of that in good conscience.

One of the things that I learned from Ed's passing was the importance of finding a qualified replacement to carry on one's own work. Ed could not find anybody who was qualified to replace him as the leader of Tall Timbers. But for one reason or another he was too busy to think about, let alone train, a natural successor, and the pool of candidates he had to choose from was not that great. I have gone through something similar in my professional life. The only other person besides Mr. Stoddard I could completely trust with this approach to forestry was my brother-in-law Jimmy Greene, who worked with us in the late 1950s and early 1960s. He fell under the influence of Mr. Stoddard, just as I had, and developed a strong understanding of our system. Unfortunately, his father died in 1962 and Jimmy needed to leave us to take over the family farm, which was a large and diversified operation. We missed him immediately and never could replace him. His background, knowledge, and ability were born in a time that will never return, but he will at least continue to influence land management in his area into the future.

The founding of Tall Timbers in the early years was important to the solidification of the Stoddard-Neel Approach, and while I can be pessimistic about the direction they are going in and the perpetuation of the method, there are some hopeful signs and developments in the last decade or two. Since the mid-1990s I have been working with the Jones Center, and they have become the chief students of, and advocates for, the type of uneven-aged single-tree selection silviculture and total ecosystem management that Mr. Stoddard perfected and I have long practiced. Some of the most important remnants of longleaf exist now in national forests and on military lands, and federal managers are doing a reasonable job of managing and protecting those resources. Public management has its strengths and weaknesses, but I am now more inclined than I previously have been to see public ownership and management as a good thing. There is also a growing public awareness of both the importance and scarcity of longleaf, and the need to return some of our coastal plain lands to forests dominated by fire-managed longleaf. As the pulp and paper industry declines, there may be some fine opportunities to push coastal plain forestry in new directions.

But there are also plenty of countervailing pressures. Development, rising real estate values, and rising taxes make it difficult for conservation-minded landowners to do right by the land and still pay their bills. But perhaps the biggest challenge to getting these woodlands managed the way I would like to see them managed—beyond the scarcity of good land managers—is the long-term nature of the commitment. To create a true multiaged stand of longleaf with a healthy understory will, by definition, take several generations, depending on the condition of the land you start with. People today who commit to this sort of management must accept the fact that the true benefits of such a decision will come decades, even centuries, into the future. The natural world is resilient, and that is my greatest source of hope. But we live in a world where people are increasingly unaware of how nature works, and that is, perhaps, my greatest fear.

CHAPTER 4

The Stoddard-Neel Approach: Managing the Trees for the Forest

I learned most of the techniques, principles, and approaches of my forestry practice from Herbert Stoddard, who was a true pioneer in coming to understand how longleaf woodlands worked. The Stoddard-Neel Approach that I have practiced throughout my career has evolved somewhat, to incorporate new scientific findings and contend with new social and economic trends, but it is still recognizably the legacy of Mr. Stoddard's work. The Stoddard-Neel Approach resists easy summary, and I cannot hope to provide a simple manual of easy-to-follow steps that anyone can use. Some people get frustrated by that aspect of my approach, but it is important to understand that one of the core principles of the approach is its rejection of formulaic forestry. A healthy, functioning longleaf-grassland system is incredibly complex across space and time. To apply to it an abstract set of silvicultural goals—a certain volume of standing timber, a specific allowable cut, an average density or age-class structure—might simplify the task of management, but it will also necessarily simplify the forest, often in ways that, ironically, make it more difficult to manage.

My goal in this final chapter is not to provide a heavily scientific rendering of the Stoddard-Neel Approach. Scientists at places such as the Jones Center at Ichauway have produced some excellent publications that speak to the scientific details of the approach, and so readers inclined in that direction might want to consult some of those.[1] Rather, I want to emphasize that the Stoddard-Neel Approach is rooted in a set of principles, values, experiences, intuitions, and techniques that developed over the decades. My goal here is to explain the approach's development and main principles in a way that is easy to understand.

One of the most important lessons Herbert Stoddard helped me to understand is that land management is an art based in science. This is a critical point and one of my most important guiding principles. Scientific research has been essential to the development of the Stoddard-Neel Approach, but I do not think that science alone can distill or teach the art, the true art, of protecting and simultaneously utilizing the longleaf ecosystem—or any ecosystem for that matter—while perpetuating it into the long-term future. The art of this type of management must be based in knowledge of the local materials that you have to work with as a land manager. And it is the art of land management that our schools struggle so mightily to teach. How can I hire a fine young person with a degree in forestry from Germany or the Appalachian Mountains or New England or anywhere outside of the region, bring him to the Red Hills, and expect him automatically to understand the longleaf ecosystem? I still do not know that much about it myself, and I have been working with these forests—shaping and steering them in certain directions—for my whole adult life. I really cannot emphasize enough the importance of place-specific field experience in the making of a good land manager, but such experience is itself an increasingly scarce commodity in today's world.

I am a land manager, not a scientist, and that is an important distinction. I think I know the Red Hills landscape about as well as anybody, but mine is a working knowledge, one derived not only from objectively observing natural processes like a scientist would, but also from constantly intervening in this natural system to achieve certain ends. The art of land management, then, comes in working with the material over long periods of time, learning how it responds, recognizing the subtleties of the longleaf-grassland environment from an applied perspective, and all the while not destroying the functioning forest. Most people do not even think about management being an art, though most good land managers recognize that it is. Artists first have to master their craft—the materials and the technique—before they can really create the aesthetic they have in mind. To practice something like the Stoddard-Neel Approach, you also have to have an interest in learning what you do not already know. Some of that new knowledge will come from science, but some will also come from work and experience. There is no universal formula for success in the art of land management; again, there is only basic knowledge, careful experimentation, adjustment, and what you come to know as beautiful and right.

Mr. Stoddard once wrote that most "university scientists are conspicuously lacking in real woodsmanship and all too apt to consider it unimportant as well. . . . After my many years in the wildlife management field I am more than ever of the opinion that the universities will never (except by accident) turn out top wildlife men or foresters until woodsmanship is stressed and becomes a part of the curriculum."[2] He always preached woodsmanship, by which he meant not only a knowledge of the woods but also a set of techniques for working with them, and I took that to heart. And again, being a good woodsman means spending a lot of time on the land, watching and working with it.

Even though I am still learning from these forests every day, I do know from experience what looks good out there, and that aesthetic knowledge is a big part of the art of the Stoddard-Neel Approach. When I look at something and say it looks good, such a judgment is based on my knowledge of what should be there ecologically, and what the forest should look like in its healthiest condition. In many ways, it was an aesthetic appreciation of these woods that led us to begin asking deeper questions about ecological functions and processes. A practitioner of the Stoddard-Neel Approach, then, must have a developed aesthetic sense of what looks right out there in the woods. That's a big part of the art.

As the Stoddard-Neel Approach developed over the last half century, we came to realize that one of our explicit goals was to manage forests so as to perpetuate as many of the ecosystem's components as possible. One thing about ecosystem management—and I consider Stoddard-Neel to be a form of ecosystem management—is that whatever you do out there is going to have cascading effects. The same is true when you choose not to do something. Mr. Stoddard would say, "Why destroy critical functioning parts of an ecosystem just because you want more money, whether it is from augmented game populations or from the trees?" It does not have to be that way. Thus, the underlying forestry principle of the Stoddard-Neel Approach is to take a modest timber harvest while protecting—and in some cases enhancing—the ecological value of the remaining resources. The Stoddard-Neel Approach is a silvicultural method primarily concerned with biodiversity preservation and protecting the total resource; timber production, while important, is secondary. That is one of the most important principles that distinguishes our approach from other even- and uneven-aged forestry methods. Some people think I cut too conservatively, but we learned through long experience what we could take without disrupting the nontimber resources. And I can guarantee you that other approaches that take a greater percentage of the standing timber when they cut either do not fully account for biodiversity protection or they do not prioritize it. As long as landowners do not need or demand the maximum financial return from their land, particularly in the short term, and they can have some patience in how they manage their property, we can show them how to have their cake and eat it too.

Mr. Stoddard started his work on land that was dedicated to quail management and quail hunting, and he had the great luxury of being able to work with some fine examples of the diverse longleaf-grassland systems that historically dominated not only the Red Hills, but much of the upland southern coastal plain. Mr. Stoddard's approach to managing land, then, began with the core question of how land and particular land-use practices produced quail. In some ways, that question was a fairly narrow one, particularly compared to the broad concerns that guide my land management today, but quail management opened up the workings of the larger system to Mr. Stoddard's view in some fascinating ways.

In the years before World War II, Mr. Stoddard managed properties in the Red Hills for quail abundance, but he never aimed to maximize quail numbers if that detracted from the larger aesthetic experience of the hunt. Mr. Stoddard's management had to maintain the classic quail hunt, which meant beautiful open woods and grasslands with all sorts of aesthetic diversity, all creating a pleasant landscape experience. I want to be clear that when I refer to "open woods," I mean woods that are visually open and present that classic long view. I am not referring to the minimal stocking of timber that some plantations have turned to as a way of maximizing quail. The aesthetic of open but well-stocked woods is still very much at the core of what we aim to produce through the Stoddard-Neel Approach—it informs the art. In fact, to a great extent it *is* the art. But maintaining that wooded aesthetic is not entirely compatible with the maximization of quail, which thrive in areas of agricultural disturbance. Unfortunately, today's owners and managers of quail lands are often more interested in maximizing coveys to the detriment of the health, integrity, and beauty of the forest. Back when I first went to work with Mr. Stoddard, you rarely heard folks bragging about the number of quail they shot or the number of coveys they flushed. There was no such thing as a standard number of coveys every day on every course, as some people have come to expect today. It did not work that way. Landowners had a more expansive notion of the hunting experience. If they bragged about anything, it was the beauty of the woods and the quality of the hunt. Hunting quail in the Red Hills was about experiencing the land as a whole.

A good hunting experience, as I came to understand it, varied from property to property based on specific quail management factors such as the availability of food and cover. And landowners back then prided themselves on the distinctive aspects of their properties and the quality of the courses they were able to lay out through their woods and fields. In those days you could go out and hunt for three hours in the morning and three hours in the afternoon; that was about the extent of the actual hunt in most cases. Some people hunted more, some people hunted less. In three hours, it was a good hunt if you could find twelve to fifteen coveys. That does not mean you shot every covey, but with a good slow hunt and good dogs, you shot an awful lot of them.

Diversity of vegetative cover, land use, and overall layout were also part of the art of land management on the quail plantations. "Natural beauty" was prized, and those who hunted for such beauty sought it both in the native vegetation and in certain human uses of the land. As time went by, good managers would eliminate some fields and add others based on factors such as soils and terrain, all with the aim of achieving the most pleasing visual experience as the hunt proceeded. All the while, they also managed the trees and ground cover to be in the optimum condition for the *continuous* production of quail, timber, and natural ecological diversity.

The owners and guests on the plantations often did not recognize the success of this art of management, but to us managers it was obvious in the discussions after the hunt. The most complimentary statements regarded the "beauty" of the woods and fields and, of course, the numbers of quail and the dog work. A hunt on an abused or poorly managed course, even when there might have been good success in terms of quail numbers, was never appreciated to the extent that an ecologically healthy area was. Even the inexperienced hunters subconsciously recognized these aesthetics, and thus the ecological values, without being that knowledgeable about the resource.

In developing and managing quail plantations, Mr. Stoddard and the people he worked for put an emphasis on the forest for several reasons. For one, they preferred hunting in woods that were visually open. In the Red Hills, quail hunting on upland without pine trees is not as desirable as hunting in a pine forest. You can hunt around fields, and that is a different landscape, but in the woodland landscape you need to have pine trees where they belong to achieve a good hunting experience. The people who owned these places valued their longleaf forests for their beauty, for what they added to the hunt's overall experience, and today the Stoddard-Neel Approach can work for landowners who are willing to defer timber returns in exchange for beautiful forested land.

But these forests were not only important to the aesthetic of the hunt; they also became an essential part of the management system that Mr. Stoddard developed. Without the forest being in a certain condition, with an open canopy structure, it was very difficult to use our most basic management tool: fire.

As Mr. Stoddard noted in his pioneering study *The Bobwhite Quail* (1931), frequent fire was critical to perpetuating the longleaf system and its capacity to produce quail. The bobwhite quail thrives in edge habitat, essentially making it a farm bird, but Mr. Stoddard discovered early on that frequent fire in the longleaf forest made ideal woodland quail habitat as well—mostly by favoring the early successional vegetation that quail favor as a food source. To increase quail numbers, the larger areas of longleaf were improved by adding small fields and thickets to create landscape diversity and cover. Fire took care of the rest. It suppressed any midstory underbrush, leaving a low understory of grasses and legumes beneath the open longleaf canopy. Without fire, what we have come to value about the longleaf-grassland system—the incredible biological diversity as well as the aesthetic experience—would disappear. Thus, frequent fire was a key contributor to the system and its particular beauty.

More than that, though, Mr. Stoddard came to realize that fire was an ideal management tool for keeping down the rough—that successional midstory of hardwoods that comes in and chokes out the understory in fire's absence. Compared with the alternatives, such as mechanical brush cutting or chemical hardwood control, fire is inexpensive, requires little labor, and is good for the forest. Moreover, under the right conditions and in knowledgeable hands, fire is relatively easy to control. But for fire management to work well, you need a certain forest structure. An overly dense stand of pine can shade out the understory, depleting the fuel supply as well as suffocating plant diversity. On the other hand, land too sparsely populated with pine trees means uneven needle fall, which in turn deprives a potential fire of this ideal fuel. One of the key goals of the Stoddard-Neel Approach, then, is to create a forest whose canopy structure will support the long-term use of fire as an effective, inexpensive management tool.

The easiest way for me to describe the historical importance of fire to the landscape in which I work is to simply say that almost everything in the longleaf-grassland ecosystem has some adaptation to fire. That adaptation has developed over a long period of time, so if you take fire out, you are upsetting the conditions those plants and animals evolved under, which means they will likely decline in abundance or be lost. I have seen this happen many times; you stop burning and the ground cover diversity disappears. So do many of the birds, mammals, reptiles, amphibians, and insects that live in the longleaf ecosystem and make it distinctive. Fire, to a great extent, is the producer of plant diversity in the system, and the same is true of animal diversity. A great example is the gopher tortoise, which does very well in frequently burned longleaf forests but struggles in the absence of burning. The gopher tortoise is a keystone species, which means that it helps hold together the larger picture of faunal biodiversity. It digs burrows all over the forest, and literally hundreds of species of animals make some use of them. They also function as refugia during fires. So if you stop burning and subsequently reduce your gopher tortoise populations, declines in vertebrate and invertebrate species will almost certainly follow.

The natural succession dynamic in the longleaf region, absent frequent fire, is for longleaf forests to turn into upland hardwood forests, oak-beech-magnolia down here in the Red Hills. The sandier soils that support longleaf in many other parts of the coastal plain would probably move toward a different species complex. But in the absence of fire, the plants and animals of the longleaf-grassland ecosystem gradually lose their ability to survive until they are gone, and then they are replaced by other species. Pockets of hardwood succession are not necessarily a bad thing. They support certain forms of wildlife and provide some aesthetic diversity as well. We have some upland hammocks here in the Thomasville area that are old growth. These oak-beech-magnolia hammocks on the hills are remnants of the patchiness of the region's fire history. They are beautiful places, and I value them. But I am not sure that I would want to stop fire and convert everything to oak-beech-magnolia. Historically, the fire-maintained longleaf forest dominated the uplands. The natural history of the region would have produced those hammocks in some places, but we do not want them everywhere, or else we will lose a forest perhaps best described as a beautiful outcome of natural, human, and evolutionary history.

It is thus to the long-term benefit, both ecologically and economically, for the landowner in the longleaf region to manage with fire. And quality fire management requires that you have the proper fuels. Grasses like wiregrass are terrific fuel for carrying the sorts of ground fires we use in our management. That is one of the reasons we value those lands where a wiregrass understory

FIGURE 8. Leon Neel in an oak-beech-magnolia upland hammock in 1964 in Thomas County, Georgia. Such forest communities occur under conditions of fire exclusion. Historically, such communities were relatively rare on the southern coastal plain due to the prevalence of both lightning-caused and anthropogenic fires. (From the Harold Biswell scrapbook, unprocessed Stoddard material in the Tall Timbers Research Station Archives, Tallahassee, Fla.)

remains. But just as important—particularly in the absence of a grassy understory—are the longleaf pine needles themselves. With a year or two of needle drop, the forest floor develops a good fuel base for supporting the fire intensity needed to control hardwood rough. But to have sufficient needles for fuel, you need to have a fairly consistent longleaf overstory. If landowners cut too many trees—whether out of a desire to increase quail numbers or to cash in on standing timber—they will diminish their capacity to easily and effectively manage with fire. Hardwood succession will creep into those needle gaps, and they will not burn as well. Then you will have to control them in other ways. You trade a short-term economic gain, in timber or quail numbers, for a long-term management expense. The Stoddard-Neel Approach respects the ecological and aesthetic values of the forest that fire historically produced in this region, but it also values the ways in which the structure of the forest allows us to continue to manage with fire.

Fire is neither a simple nor a single-purpose management tool in the Stoddard-Neel Approach. We use fire for specific purposes, and that is where management begins to get complicated. Again, Mr. Stoddard drilled into my head that the application of fire is the land manager's art, and one of the most important things I have come to appreciate about fire is that it can be revelatory. The simple act of putting fire to a piece of land can help you to see what was there in the past, and that, in turn, can help you to make future decisions about when and how to burn. Fire is such a useful management tool because it tends to the complexities of the longleaf system. In a sense, when fire is applied and managed well, it does some of your thinking for you.

We take the scientific knowledge that has accumulated over the years very seriously, but there are multiple factors to consider in fire's application that do not happen under the controlled conditions of a laboratory. Understory species composition and uniformity of the fuel load is critical. When we burn, we look at what type of fuel there is and how much, and we pay attention to the condition of the rough. You also have to look closely at the terrain and ask whether or not the land in question is contained—you have to be able to stop a fire once you start it. And, of course, there is the weather—the wind direction, wind speed, and relative humidity can make the difference between a slow ground fire and a raging inferno. How you use fire also depends on what you are trying to accomplish. If you are caring for longleaf regeneration and reproduction, you do not want to burn too hot late in the burning season when the trees are in the candle stage—after several years of growth, the terminal bud grows out very quickly on young longleaf and looks like a candle. Longleaf pines can withstand a lot, but those candles are sensitive to fire, and so you want to burn areas with lots of regeneration a bit earlier than you do mature stands. Taking all of these factors, practical as well as ecological, into account is where the art of fire application comes into play.

Mr. Stoddard burned by common sense, not a specific set of codified criteria. He knew by experience when land would burn too hot, or when it would not burn at all. Now that does not mean that sometimes he did not test, particularly when he was trying to get a real cool fire. Following Mr. Stoddard's example, I often go out and set a fire only to find that it will not burn with the desired intensity. But I know that if I had waited two days to set it, it might have blown the roof off. Conditions can change that quickly. I would prefer to start early and keep going back until I can get the cool fire that I want, rather than make a mistake and blow up something. Good land managers are humble in the face of the lands they work with, and working with fire will make you humble really fast.

If the primary reason for owning your land is to hunt quail, as it was for many of the early conservation landowners in this region, then it stands to reason that you do not want to burn during the quail season. Today, the season usually runs from November to March, when the young birds are grown and the frost knocks back lush growth. You need to have proper ground cover during the season, and a fire would disturb that cover habitat for a time. It also stands to reason that you would not burn during the summer nesting season. Even though quail will renest, you cause all sorts of complications that might ultimately reduce quail numbers. So that is why our system developed to burn at the end of quail season, which was about March 1, and before summer nesting really got into high gear. Mr. Stoddard always said that any quail nest destroyed before June 1 was probably beneficial. He did not recommend destroying nests, but if they were casualties of a spring fire it probably did not hurt too much because the birds would renest, and the conditions were sometimes better for a larger brood after June 1. The goals of quail management, then, established the spring burn as an early foundation of Mr. Stoddard's fire management approach. Typically, we burned in the months of March and April, and occasionally a little bit later.

Over the years, the Stoddard-Neel Approach has evolved from a focus on quail abundance to a land management system with a broader focus on ecosystem health. That evolution, in combination with the discoveries of fire ecologists, has affected how we have used fire. It has only been in recent years that scientists have begun to question, and rightly so, when and how fire might have occurred in the state of nature. Lightning was the primary ignition source for the natural fires that shaped the longleaf forests of the coastal plain, and since the lightning season is in the summer, that complicates modern management. As a result, we have moved to doing more summer burning.

Scientists have pretty well proven that some plants in the longleaf understory are adapted to summer fires, with wiregrass being the classic example. Wiregrass will normally seed after a summer burn, and so its presence as a dominant part of this plant community is proof of the historical frequency of summer fires. I heard hours-long discussions among Mr. Stoddard and others half a century ago about how wiregrass may have adapted to fire, but it was pretty far down the ladder of his things to work on. We were all interested in this question, but nobody got around to it until Ron Myers, who now works with the Nature Conservancy, and others worked out the life history of wiregrass.[3] Now that we recognize the importance of wiregrass as an indicator species of a relatively undisturbed piece of land in the Red Hills, and as we have come to understand and appreciate the incredible understory diversity of such undisturbed landscapes, we have started to think more about how we time our managed fires. Of course, in this fragmented landscape of remnant longleaf patches, we can no longer rely on summer lightning to do the work for us. We now have to mimic what used to happen naturally, and, depending on the piece of land, that may mean occasional summer burns.

But we cannot get too caught up in when fire happened naturally in this system, particularly if moving to summer burning might compromise our other management goals. The longleaf landscape I have come to know is a relatively young environment, only several thousand years old, and there have been people and fire in these woods for that entire time. In other words, defining and mimicking what "naturally" occurred in this system is not an easy thing to do. Indians burned the land for thousands of years before we showed up, though we can only guess when they preferred to burn and for what purposes. But it seems likely that they burned not only during the summertime but also during the spring, and perhaps at other times as well. The forest as we know it thus took shape in concert with human action, and it is difficult to say whether natural or anthropogenic fire had more to do with the system's development. Such contingency requires that we pay close attention to how flora and fauna react to fire in all seasons, and that we tailor the application of fire in a way that makes those ecological reactions meet our specific management goals. A fire in one season will do different things than a fire in another season, and we are just beginning to figure out the larger managerial implications of that.

As far as modern land management in the longleaf goes, your goals have to be specific when deciding on the season of fire. If you want wiregrass to seed, then you burn in the growing season. On the other hand, a lot of legumes seed better after a cool season fire, so if bringing some of that diversity out is your goal, you might opt for an early spring burn. If you are restoring a stand of timber with a ten- to twenty-year rough, you need to start with cool winter fires. A hot summer fire can kill your forest in those conditions. On the other hand, warm season fire tends to hit brush pretty hard, so under the right conditions experienced burners might want to blow a hot fire through an overgrown piece of land, though they had better know what they are doing. There are, in other words, no hard and fast rules about when and how to burn. Under the Stoddard-Neel system, your burning decisions will depend on your management goals and the specific landscape to which you are applying fire.

The longleaf pine environment, like all environments, is dynamic, and the Stoddard-Neel Approach has to be dynamic as well. Herbert Stoddard stood firmly by his work, and the field knowledge behind it, but he did not hold fast in the face of contrary research. The same goes for my approach, which has responded in recent years to this new research on the historical timing of fire. The scientists, in other words, can tell us a lot about how plants and animals evolved with and adapted to fire in this system, and I value that. But few of them know fire as a management tool, when and how it should be used and to what effect in the broader landscape.

I do not talk like a scientist because I am not a scientist. I know a lot of scientists; you ask them about the season of fire, and they will give you a very learned academic answer. But in this age of burning restrictions and permits, when your chances to burn can be few and far between, land managers rarely have the luxury of scheduling fire. Instead, we have to take our opportunities when they come. And, from a land management perspective, I can say with great certainty that the *frequency* of fire is more important than the season of fire in protecting and restoring the larger longleaf system. If there is one hard and fast fire management rule in the Stoddard-Neel Approach, it is to burn frequently. That is especially true on good soils, which will promote rapid understory growth. If you let a longleaf forest in the Red Hills go three, four, or five years without burning, you will start seeing some long-term ecological effects, mainly in the form of encroaching hardwood species. I do know that the more we burn the woods, the easier the woods are to maintain, and the easier they are to manage in terms of both longleaf reproduction and biodiversity. But if you wait for a particular moment to burn, and then the conditions either prevent you from doing so or lead you into a poor decision, you make your job more difficult the following year and into the future.

In the absence of fire, you not only risk a thicker rough, but your fuels settle and a lot of the plants are mulched out; that is, pine needles and debris from neighboring vegetation get matted down and smother out the native ground cover. Under such conditions, nothing can reach the mineral soil except an armadillo. Moreover, we have annuals out there as well as perennials, and fire affects them differently. So fire management is like a big puzzle. There might be a way to create a broad management formula for using fire in the longleaf region, but each acre of land has its own potential as well as its own set of problems. If you get too formulaic in your burning approach, you will miss that specificity.

Anyone using fire must understand that it can take on different personalities. Fire should never be used in a way that abuses the land, and you have to know enough to know when it is abusive and when it is not. Fires are not always productive. I have seen a lot of people burn way too hot, for instance. There may be times when you need a hot fire. In reclaiming old-field land, or even virgin land that has been allowed to grow up in the early stages of hardwood, you ideally need a hot fire. It is surprising what our pine trees will take. Nonetheless, there are certain times when they are more vulnerable to an intense fire. As I have already noted, longleaf reproduction is vulnerable at several feet high when it puts up new growth in the spring. The heat blast itself—meaning just the increase in temperature—can defoliate, killing the buds and sometimes the trees too. So when you are trying to encourage regeneration in an area, you probably want to avoid a hot summer fire. But mature longleaf is pretty hard to kill with properly set fire.

Again, you just have to know the personality of fire under different circumstances. If it is hot, dry, and windy, you will get a hotter fire with the same fuel load than you would when there is no wind and higher humidity. Everybody will make a mistake on occasion, but the way to avoid a catastrophe like that is to burn frequently. Burning frequently makes it easier to use fire with precision, and a frequently burned landscape is more forgiving of mistakes.

One thing you have to pay attention to when you burn is the fuel load itself. From a practical standpoint, in our area we have two main sources of fuel. Other longleaf areas have different fuel sources and they probably burn differently because of it. One fuel source anywhere in the coastal plain is the pine needles, and of course longleaf needles are the best fuel source of all the pine trees. They are longer and they do not become compact when they fall to ground. They are also more resinous than other pine species. Slash pine is the next best among the pines, then loblolly and shortleaf in that order. The second source of fuel is the ground cover. Now, if you have a herbaceous ground cover—it does not have to be wiregrass—it would normally include a lot of grass species, plus the other plants in there, whatever they might be. Even some hardwoods like *Quercus pumila*, the runner oak, evolved under fire. All of that is highly combustible, too. So if you get that mixture of herbaceous ground cover, which is only a foot or two high in most cases, interlaced with the falling needles over a year or two, then you have a joined fuel load with plenty of oxygen, and that burns fine. But if the ground cover gets mulched down over time and the bottom layer begins to rot, it can hold too much moisture and be hard to burn.

There are times when we need to force fire into places that do not normally want to burn. There are several classic cases depending on the region. Down here in the Thomasville region we might have a property with several hundred acres of beautiful upland longleaf forest on our slightly rolling hills, but then we might get a drain coming through at the bottom of the slope. Our soils carry more water than the sandier soils in the Albany region, for instance, which is part of the Dougherty Plain. Water seeps into the ground more quickly up there. But the drains in the Red Hills are true hardwood drains with flowing water, and they constitute different ecological communities than the upland pine forest. They add tremendous diversity to our longleaf forest. On either side of the drain there is an intermediate area that ecologists call an ecotone, and that is where our bogs occur, depending on the terrain. Those bog and creek areas will sometimes burn out under dry conditions if we have a few pine trees along the edge to provide some fuel. But the hills will always burn, and the hills will burn quicker and hotter than the creeks. So we burn the hills first if we want to get fire in the bottoms. That is just common sense. You burn your hills out early and then wait for a hot, dry day for burning and burn through that creek. That way you do not have to worry about the fire getting away from you into the unburned uplands.

It is a good idea to get fire through those drains on a certain frequency, because those hardwood communities are constantly working to expand. The plants in the creek bottom are trying to move out and expand the hardwood site. Fire is the easiest way to keep such encroachment down. Historically, the uplands burned frequently and easily, with fires usually petering out on downward slopes toward the lowlands, depending on moisture levels. But we know that these drains would have burned during drought years in the past, so when we put fire into them during dry times today, we are mimicking the natural and historical variability of fire ecology in relation to climate.

In today's world, we face several problems when it comes to the application of fire in the Southeast. The West gets most of the publicity for bad forest fires, and it is well deserved. They have a tremendous job ahead of them to restore fire into those landscapes. But even in the South we have had bad fires that have gotten out of control and destroyed property. Many result from not burning regularly. If we burn when we can, even if it costs a little money, it is not going to cost near as much as a catastrophic fire, which results from neglecting the region's fire ecology. I remember watching the

Okefenokee fires of southeast Georgia and northeast Florida that burned throughout 1954, 1955, and part of 1956. They were not unlike the recent Okefenokee fires of 2007. I went over there with Mr. Stoddard several times to watch those fires go. We saw fires jump two- and three-hundred-foot rights-of-way where the railroad and highway were close together. We were in the middle of a severe three-year drought, but we did not have any problems in the Red Hills. The quail plantations had already been burning regularly for many years. For instance, on the home place at Greenwood, which has a highly combustible ground cover with a heavy stand of longleaf and wiregrass, we simply switched from day burning to night burning to accommodate the drought. Because of calmer winds and higher humidity, a fire will burn much cooler at night, so we were able to keep right on burning through the drought. It so happened that the drought was followed by two years of good longleaf seed mast, and, to regenerate naturally, longleaf seeds need bare mineral soil exposed by fire. So we got a lot of reproduction out of those drought-year fires. If those lands had not been regularly burned, and the drought had forced us to put off our burning, we would have missed those mast years. Regular burning not only ensures diversity, but it is also a hedge against extreme weather.

Another problem we face in fire management is development and the fragmented nature of today's forested landscape. In the past, everybody around here was psychologically geared to burn in March and April, so once that time rolled around, we burned. On most of the quail plantations, the manager of the property directed everything. He knew the property, and the woods burners on most of the places were experienced. They knew the land and how fire would behave on it, so the risks were minimal. When conditions became ideal, the manager would send the crews out and they would set fire to the place. Fire management was easy then, because we had complete freedom in choosing when to burn to suit our particular management needs.

We do not have that same freedom anymore. Things are getting worse, not because the fires are worse but because more people are in the path of fire and smoke. Today, with the rules and regulations, I sit here sometimes letting good burning day after good burning day go by because I cannot get a permit. Weather conditions may be such that the smoke will not disperse straight up. It hovers close to the ground, and we do not want smoke on the highway. Moreover, the public increasingly complains about smoke, particularly when it descends on populated areas. Or, alternately, there may be a day when conditions are ideal for a good hot fire that would help a piece of land, but I cannot get a permit because it is too dry and the authorities are afraid a fire might get out of control. The Georgia Forestry Commission is doing what they have to do. They understand that we need to burn and want to burn, and they cooperate with landowners very well. But regulations still take probably a third of the burning days away from us, which is a major hindrance. And I am afraid the restrictions will get worse before they get any better.

Given these restrictions and limitations, you begin to see how academic these debates about the proper season for burning can be. Those of us who actually do the burning have learned that we have to do it when we can. We burn when we get our best chances and the most suitable conditions, under the principle that frequency is the key to maintaining the system. We wish we had more latitude to burn with many different types of fire at different times of year, but under the circumstances we burn when we can. The real world out there is not a controlled laboratory.

Because canopy composition is critical to using fire as a management tool in the longleaf region, the Stoddard-Neel Approach hinges on managing trees over time to achieve and perpetuate optimal forest structure. In the early years Mr. Stoddard cut very little timber from the Red Hills quail plantations. He did begin advising preserve owners to thin timber during the late 1920s, but it was not until World War II that they considered harvesting on a larger scale. Not only was the timber market on the upswing with the war, but the longleaf and other pine stands on many of the plantations were becoming dense with timber, thus choking out important quail habitat and lessening the land's ecological diversity. World War II, then, represented an important confluence of events that gave birth to the timber management focus of the Stoddard-Neel Approach.

There were several factors involved in Mr. Stoddard's decision to broaden quail management to include forest management in the decade or so after the quail study appeared. First of all, he had to make a living. The Cooperative Quail Study Association suspended activity in 1942 and folded the next year, so his primary income source was dwindling. The landowners trusted him and he knew those woods better than anybody, so they had a definite interest in keeping him around. Mr. Stoddard had told the owners for years that they needed to thin some timber stands that were too dense for quail and other wildlife, but it was really World War II that got them thinking seriously about harvesting some timber. The government, of course, wanted all the lumber they could get during the war years. They were encouraging landowners around the country to cut their timber so they could have the lumber necessary for the war effort. There were a few sawmills in the area, and they were trying to find good timber to cut. They had talked some of the preserve owners into cutting some timber, but the results were truly atrocious. The worst land that I manage today, as far as restoration is concerned, is land that was cut by people who overcut their timber back in the early 1940s. So Mr. Stoddard saw the opportunity not only to make a living, but to do the job right. He came in to represent the landowner in these timber transactions, and in some ways he became a regulatory presence over the timber cutters.

When Mr. Stoddard began moving into forestry, there was a smaller total volume of timber on the lands he managed than there would be by, say, the 1980s, when the plantation lands were still pretty well held together. On most of the properties the forests were younger during the interwar period, and over the years we developed a system that encouraged growth. We promoted a system where we were cutting more timber as the years went by, though we were also growing more timber. World War II, then, was a moment when the forests of the Red Hills could withstand a bit more cutting, and such cutting in turn led to a deeper inquiry about how timber management could be used sustainably to produce a steady stream of income and to perpetuate the ecological integrity of the Red Hills.

In the years before I joined him, Mr. Stoddard had been building timber capital on the lands he managed, and so when he began cutting more timber during the 1940s, he thought a lot about how best to balance growth with harvesting. Out of that experience came another core principle of the Stoddard-Neel Approach: treat your forest like an annuity, harvesting a *portion* of your annual interest rather than all of it. You certainly did not want to spend down your capital. In fact, under the Stoddard-Neel Approach, we aim to put more money in the bank every year through timber growth than we take out through timber harvesting. That makes our approach different from other methods of sustainable timber management, where annual harvests are theoretically pegged to match annual growth. Our approach is more conservative, largely for ecological reasons. Rather than maintaining a constant timber volume, we like to see our investment grow over time. Such a conservative approach gives a land manager maximum flexibility in achieving noneconomic goals.

The primary reasons Mr. Stoddard wanted to start cutting some timber on the places he managed was to maintain consistency in the forest canopy and to create a multiaged forest. This goal of multiaged stands of timber is another important difference between Stoddard-Neel and conventional commercial timber management approaches that rely on single-age-class pine plantations and clear cutting. Our approach is best for those landowners, private and public, who want to generate sustainable long-term income while also protecting, and in many cases enhancing, the ecological integrity of a piece of land. Our approach is innovative exactly in the way it can balance commercial use and landscape preservation.

FIGURE 9. Herbert Stoddard began his forestry work in 1941, after a hurricane damaged many longleaf woodlands in the Red Hills and demand for wood products escalated during World War II. His colleague and friend Wallace Grange took this photograph of him next to an old-growth longleaf pine around 1950.

In terms of sheer efficiency of cellulose production, the Stoddard-Neel Approach cannot compete with the pine factories that have taken over so much of the South. Most of them grow trees—generally slash and loblolly pine—on a short rotation, clear-cut-and-plant model that is effective for their purposes. Such systems work well if your only goal is to maximize wood production—to farm trees. But such a system is input intensive, it protects few ecological values, and, in cases where good forestlands have been cleared to create pine plantations, it has also done substantial ecological harm. Plantation pine is a monoculture, plain and simple. There is, however, some hope. The pulp industry has been pulling out of the South for the last decade or so, and while that makes it difficult to sell pulpwood, it also makes growing for sawtimber more attractive, which is what the Stoddard-Neel Approach is geared toward.

For decades, monocultural pine plantations relied on slash and loblolly pine because they were easy to plant and they seemed to reach harvest size more quickly than longleaf. In fact, while longleaf can be slow growing in its early years, it can catch up to these other planted pines later in life, even within the temporal confines of fairly short rotations. Moreover, it is now also possible to plant or in some cases seed in even-aged longleaf reproduction. Having longleaf instead of these other species is generally an improvement. But again, from both an ecological and fire-management standpoint, I find these systems lacking. Seed-tree and shelterwood systems of silviculture tend to have minimal age-class diversity, with a dominant cohort of even-aged trees and a second cohort of older seed trees. They may produce trees more efficiently than the Stoddard-Neel Approach can, and in a slightly more natural way than pine plantations do, but their aim is still to maximize wood production at the expense of other ecosystem values.

There are also other multiaged timber management systems out there, and while they are a bit better by the standards we use to measure success, they too have critical shortcomings. One uneven-aged approach is group selection, which usually involves creating big rotational gaps in a larger forest, such that your age-class diversity exists in clumps. This approach does allow for natural regeneration, but it also results in a forest that diverges dramatically from the structural diversity of natural longleaf systems, and the big gaps it creates can produce all sorts of fire-management problems because of insufficient fuels in these gaps and the hardwood succession that can result. In many ways, group selection is like clear-cutting on a smaller rotational scale. From an ecological standpoint, the disturbances created by group selection tend to be too large and sudden to maintain consistent fire management.

Some observers argue that the closest silvicultural approach to Stoddard-Neel in terms of its small-scale canopy disturbance is the BDq approach. The BDq system attempts to mimic natural disturbance regimes to a certain extent, but it also is formulaic in how it goes about doing so. Moreover, its goal is the maximum sustainable harvest, or taking 100 percent of the growth increment, which will necessarily mean incurring certain ecological costs. For instance, the "D" variable stands for a maximum tree diameter under this system, which means once trees reach a predetermined size and age (and these are not always the same thing, by the way), they are automatically cut. In effect, the forest has a maximum age beyond which no tree can grow. As a result, there is not a true diversity of age classes, and there is little place in this system for old-growth trees or any decadence. Timber selection and harvesting is also based on a standardized residual basal area formula (the "B" variable), which means that harvests are pegged to the abstract goal of an ideal standing timber volume. And the formula by which they seek to distribute age classes through the forest (the "q" variable) is also too artificial as far as I am concerned. While this system is certainly preferable to the other ones I have described, the use of abstract formulas is an unnecessary and costly surrogate for the careful single-tree selection that is the hallmark of the Stoddard-Neel Approach.

In our multiaged silvicultural system we do not want the same consistency of age classes throughout the forest. Diversity is what we want. We might want a thick stand of pine trees here, an intermediate stand there, and a sparser stand in other places, all with different age classes. Moreover, in a multiaged forest, every acre does not have what we would call a forest on it, at least in an ecological management system like ours. You might have an acre with no trees on it right now, but ten years from now it is going to be full of seedlings. And thirty years from now it will have some nice trees on it. Such small open gaps are where you generate your future forest. In the meantime, one of these areas that once had a lot of trees on it will not have any, or maybe just a few, some time in the future. There is constant regeneration, growth, maturation, death, and decay. That is the way it works in nature, and in our management system we try to keep it that way.

This is where our harvesting approach really comes into play. If one's goal is a diverse, multiaged stand—and that is what the Stoddard-Neel Approach aims to achieve—ecological management in the longleaf pine–grassland system often requires that we take some trees. The longleaf biome developed with disturbance as a key defining feature. Fire was one disturbance regime that historically affected these lands, but windfalls, insect infestations, disease, and other sources of mortality worked to create openings in these forests that we try to mimic when we take timber. Indeed, we are coming to understand that the disturbance history on a particular piece of land is critical to its diversity. Take away a careful, controlled harvest, then, and we might lose the biodiversity that the land's disturbance history created. Of course, it has been demonstrated time and again that harvesting can also do a great deal of damage, so we are very careful in making our selection and removing the trees. But in a system defined by disturbance, a moderate amount of timber cutting can be a good thing from an ecological perspective.

The hardest part of the Stoddard-Neel system is going out and marking timber, and making the right selections, because whatever you take is going to be gone once you take it. And that raises another critical feature of the Stoddard-Neel Approach: ours is a single-tree selection method, and ecological considerations are paramount in the decisions we make about harvesting trees. Mr. Stoddard always said, "The trees you leave are more important than the trees you take," and I have tried to follow that advice. One thing that he impressed on me was that he treated the timber stands in perpetuity. You have the forest as an entity, and then you have each individual tree as an entity. At some point every tree in a stand over a period of time will be replaced by another tree from the ground up. In other words, when you are harvesting trees you are also establishing reproduction to replace those trees. The aim of marking and cutting timber is to perpetuate the forest—that might sound counterintuitive, but it is true. As a result, good marking and cutting demands a lot of imagination and forward thinking.

Creating and then perpetuating a diverse, multiaged forest is the principle behind how we mark and cut trees under the Stoddard-Neel Approach. But we do not have a formula for marking timber, such as achieving an ideal basal area or taking a certain number of trees per acre. And that makes this a difficult system to learn.

When I first started out, all I did was tally. Mr. Stoddard would not let me touch a paint gun. I did not know how to make a selection, so I would tally the volume for him. We would tally every tree individually. He used a diameter stick, a manufactured scale stick that you could hold at arm's length up to the tree to get the diameter at breast height (dbh), which is the tree's diameter at four and a half feet off the ground. With the stick he had, you could convert the dbh into an estimate of the board-foot volume of the tree, based on the number of logs that you estimated. So we got a 100-percent-volume tally without having to use the volume tables most other foresters use. From a practical standpoint, this system worked well. He had tested that stick for accuracy, but there is no such thing as a totally accurate scale in timber. There is a human element involved all the way down the line, so tallying and marking timber is always going to be an inexact science. We were just trying to get the most accurate estimate of the volume that we could, and then it was up to the potential buyers to determine what they thought was there.

We always used white paint to mark trees, and we always marked on the same side of the tree, so that others would not have to hunt for our marks. When you get into a thick stand of young sawtimber and you put all the marks on the same side of the trees at eye level, it sometimes looks like a solid white line. On several occasions we have had landowners get extremely upset because they thought we were marking too much timber. So I managed to prevail on them to first walk in the woods with me and let me point the trees out, but that did not always change their minds. The white paint gave them a false impression that we were about to clear their forest. On a couple of occasions I asked them to let me get the saw man to go ahead and cut some trees down while they watched so they could see what was left. They were always amazed that only a few trees came out, and it improved the aesthetics so much. We took our marking very seriously, and Mr. Stoddard taught me from the beginning that there is no perfect way to mark timber. There was one thing, however, that Mr. Stoddard was always clear about: he never marked a tree just to add volume to a sale. He took the timber he thought he could take without harming the present and future state of the forest. Our marking was also guided by the fact that we had to look ahead and anticipate what might be removed in the next cut, or the cutting cycle after that. In the Stoddard-Neel Approach, one cut is not the driving force.

Most professional foresters get frustrated when they hear me talk about marking and cutting timber. There is no abstract formula for marking timber in the Stoddard-Neel Approach, and they cannot seem to wrap their heads around that. We do not use basal area to determine an allowable cut. Nor do we use dbh to tell us what size trees we want to mark. In other words, we do not have a standardized target stand structure to guide the marking process. I just think the forest deserves more attention than a formula can provide. Every time paint hits a tree under Stoddard-Neel, it is the result of serious thought about the ecological effects of cutting that particular tree. Will the resulting gap be too small for longleaf regeneration? Will it be too big to carry a fire? How will the trees around it respond to its absence? What will it mean for the landscape's parklike aesthetic? Will it contribute toward creating or maintaining biodiversity? If we take this tree this year, what will we have to do on this spot next year, or ten years from now? We do not even have a set list of questions; instead, we understand from the beginning that each stand of timber will demand its own set of questions. Each stand is unique, and each individual tree deserves a new deliberation over what will happen if we take it.

There are, nonetheless, several guidelines involved in marking trees using the Stoddard-Neel Approach. Our most basic rule is that we usually take the poorest-quality trees and leave the best-quality trees. That increases the value of the trees remaining, so that, in the long run, you do not have to cut as many trees to yield the same amount of volume. There is no high-grading, or cutting out the best timber, in our system. It takes two inferior trees to equal the value of one superior tree, so your early harvests may not be particularly lucrative. But selecting to remove the poor-quality trees means that, as time goes on, you will have a forest of increasingly valuable trees, and you will not have to liquidate the stand to get a good return. Over the short term, the result may be that you get less revenue, and critics of our system have pointed that out. However, I can show them that, in a sense, we have much more money in the bank than they do. And we usually have much healthier ecological conditions as well.

From a forestry standpoint, then, our system favors the harvesting of inferior trees, particularly when we are getting a forest in shape. But, from an ecological standpoint, you cannot always stick to that rule. For instance, in many cases it is the inferior trees that contribute to the diversity of the ecosystem. If every tree were a perfect one, you would not have as much life in the forest as you would when you have some defective trees. Thus, we always leave some defective and dead trees, and sometimes we choose a superior tree over an inferior one for ecological reasons. Remember, too, we have never logged an entire plantation at one time. We usually only mark a couple hundred acres at the most on a ten-thousand-acre piece of land, so even if we take a few of those inferior trees off of that small plot, there are always plenty left elsewhere to add to the land's biodiversity. So there is a balancing act involved. In fact, everything about this system is a balancing act. That is one thing that makes it so difficult to follow through with year after year. Most people do not care to take the time and effort to mark each tree individually, or to remember what they did on a piece of land the last time they marked it. We spent enough mental effort on these tracts of timber that we knew the land and we always knew what we had done before. We were trying to create a picture, and we remembered the trees that we marked and trees that we left. And we were always thinking several cuts into the future.

Although there is no formula that adequately renders our approach, we worked out some numbers to help with our allowable cut. Our timber cuts, like most others, are based on a percentage of the total growth on the property. Mr. Stoddard, when I joined him, felt like he could take 90 percent of the annual increment and leave 10 percent of the growth to build on, along with the total starting timber volume. That was a bit less than the standard sustainable forestry formula, which takes 100 percent of the yearly growth. He was starting with a surplus of older trees that had not been culled, and so it was easy to find lots of trees that were, from an ecological and management perspective, good candidates for cutting. I worked with that figure of 90 percent for a while, but it became clear that we were overcutting from the standpoint of keeping the diversity of age classes out there. You cannot really sustain an income from timber over the long haul by taking 90 percent of the growth, unless you have a perfectly distributed age-class system on the total property to begin with, and that was, and still is, rarely the case. If you get down to where you have a healthy little tree and a healthy big tree, you normally take the healthy big tree in a long-term program. But under ecological management, you cannot apply that rule 100 percent of the time, because you want to leave old trees, both for the ecological services they perform and for the beauty they lend to the forest.

Thus, another major principle of the Stoddard-Neel Approach is that we want to keep some of the trees in the forest standing well beyond their productive timber-growing years. From a commercial forestry standpoint, this is inefficient, but from an ecological and aesthetic standpoint, it is critical to what we are trying to achieve. Aesthetically, a big tree is simply more outstanding, and, ecologically, older trees are essential for some animal species. The best example is the red-cockaded woodpecker, which builds its nests in living pine trees so that the running sap provides a defense mechanism around the cavity hole. Red-cockaded woodpeckers prefer older trees because they often have the red heart rot at the center of the trunk that will make the arduous job of excavating a living tree that much easier. Not surprisingly, red-cockaded woodpeckers are endangered today because we have so little mature pine in an otherwise well-forested landscape. So if we want to bring them back, we need to build up a population of older trees. Moreover, if a red-cockaded woodpecker ever abandons a cavity, that becomes the Waldorf Astoria of the forest. All sorts of other animals will use it. Scientists have identified a couple of dozen animal species that will take advantage of these tree holes. Also, any scar in an old-growth tree creates habitat for reptiles and amphibians—skinks, lizards, snakes, and more. And eagles like to nest in the tallest, biggest, oldest trees in the forest, particularly those old longleaf pines that develop what we call a flat top. Longleaf pines can live to be four hundred years old or more, although, like most of us, their growth slows about a quarter of the way through their life span. We strive to create an all-aged stand to accommodate as much natural diversity as possible.

To move some of the standing trees toward old growth status, I have had to come down in terms of the percentage of growth I feel comfortable taking from a forest in any given cut. First I went from 90 percent to 80 percent, then to 60, and now I am down to 50 percent of the growth. I am not as worried about the volume increase as some foresters would be. They are concerned about growth slowing down and yields declining. I am worried about maintaining a diversity of age classes out there, and the ecological diversity that comes with it.

To cut timber sustainably while also promoting a diversity of age classes, we have to gradually convert the big trees to little trees. You might need ten little trees to equal one big tree in volume, so we are slowly shifting what we take to get the volume we need. We are converting older stands to younger age classes all the time, but we do not just cut every tree over a certain dbh to do it. It is important now more than ever to practice ecological forestry on a multi-aged basis, and not just straight economic forestry. If you are not cognizant of the ecology of the forest, you are going to eliminate all sorts of habitat niches without altering the volume at all. Many foresters will stick by their dbh formula no matter how it affects the ecology, but that is not what we do. In the best of all worlds, if we had a wide total distribution of age classes and a maximum stocking of timber on all of our lands, then we could come closer to pursuing a formula for our annual allowable cut. But the environment does not work that way. Each stand of timber is different, and we have to mark each stand according to what it tells us. You simply cannot cut a big volume of timber for a lot of money year after year, because it will not last that long. And, on top of that, you will risk losing your fire management capability.

The longleaf forest makes its own demands on the practice of silviculture. At one time, for instance, some of the national forests in the longleaf belt were working on a similar financial basis in terms of not taking the total growth increment. But they were doing it by clear-cutting blocks, which will severely challenge your ability to use fire to control the rough. Even though they might be taking 50 percent of the increment, they certainly were not practicing the Stoddard-Neel Approach. Even a ten-acre block on a property of five thousand acres is, in my judgment, too big of an area to clear-cut if your goal is to practice ecological forestry within this system. The principle is that we do not want to clear an area of established woodland. We open up small gaps gradually with the help of natural events. Then we can encourage natural regeneration in the gaps. Clear-cutting disrupts the use of fire, it fragments the forest, and it, in effect, terminates a portion of the forest, which is something we never want to do.

Foresters are not the only ones who sometimes have a hard time understanding or adjusting to our approach to harvesting timber. Because many of the lands we have worked are among the most important and healthy remnants of the historic longleaf forest, we have to contend with those who do not understand how cutting trees can actually enhance ecological diversity. Some think we should simply protect these forests and leave them alone. We have to explain to them that disturbance events are critical to the longleaf-grassland system, and that is what we are doing when we burn the woods or cut trees. Longleaf needs bare mineral soil for seed to sprout, and plenty of sunlight and space in which to grow. Without those openings, regeneration will struggle to get a foothold, and you will risk the future of the forest. Over the short term, you might get a forest that, as it gets older, becomes more beautiful and diverse, but if the age class gets too old, you risk losing the entire forest to the natural life span of the trees or a major disturbance event. Management under the Stoddard-Neel Approach keeps these forests in better shape for the future, and often that management involves, even necessitates, selective cutting.

We aim, on a small scale, to re-create lightning events, blow downs, and other natural mortality with our cuts. But we do it gradually. We seldom cut a large opening in the forest, because we want to save some trees to cut the next time around. We prefer to get the reproduction first, in a small opening, and then with the next selection cut we expand the gaps where reproduction is already established. Expanding gaps through successive cuts will give reproduction more room to grow, thus slowly allowing our timber volume to increase. That works well in a longleaf-grassland ecosystem when you have virgin ground cover. If you have virgin ground cover and cut too large an opening, you are going to lose the ability to burn efficiently, increase hardwood encroachment, and then lose the virgin ground cover. Small openings are the best way to protect that ground cover.

When you do something like expand an opening to enhance growth in the next generation, you do not just go in and cut any tree. Say you want to put more light in a gap, and you have a round, quarter-acre hole with good reproduction in the middle and an even distribution of big trees all the way around it. You have seeded the area to a perfect stand of thick, young longleaf, and you want to give that reproduction more room to grow. You do not want to give it too much space because you do not want to cut trees just to be cutting trees—then you are reducing the number of trees that you can cut later on. But you do want to cut a tree to open that gap and enhance young growth. You have the same quality trees, health wise, in each direction along the edge of the gap. Which one should you take? My answer would be to take one on the south side. The sun is in the southern quadrant in the South, and removing trees on the south side of the opening allows more sunlight into the gap. By taking that one tree, you get a tremendous amount of light in there, and you are still leaving the other trees to grow and continue to build volume. That was the emphasis that Mr. Stoddard put on making a selection in the woods, and that is how I continue to work. We know that under normal circumstances we will need to go back in there in a few years and take more trees out. The object is not to spend all your money at one time. Invest it and let it grow a little bit, then you can come back and take some more trees, and it does not hurt as much. And you can still have a healthy, beautiful forest.

The Stoddard-Neel Approach has endured its share of criticisms, but perhaps the most difficult to counter deals with its economics. Our system developed on the lands of some of the wealthiest families in the nation, so early on there was no overwhelming incentive to generate capital return. The preserves' owners wanted recreation and natural beauty from their lands, and they could afford to take a loss if need be. Over the last forty years or so, though, we have had to adapt the approach to meet a rising emphasis on the financial bottom line because fewer owners can afford not to make at least some income from their lands. The Stoddard-Neel Approach is not the most lucrative of forestry methods in the short term, but I am convinced that it can meet the economic needs of landowners who care about the environment. It just takes patience and a commitment to the values that inhere in these landscapes. If an owner does not care about these ecological and aesthetic values, our system may not be right for them. But if they do, we can show them how to make a bit of money while also being good environmental stewards.

If a landowner came to me right now and asked how this system might benefit him and his land, I would first say you have to be a conservation-minded landowner. I strongly feel that my first responsibility is to the ecological health of the land, and that's how I manage. But I also understand that in today's world the cost of owning land is astronomical. Our system helps to underwrite those costs, and with patience and time it will also generate a net economic gain. But I will not liquidate a client's timber; that goes against everything I believe in. The system is set up to be adaptable. If I want to keep a client, I often can mold the system to meet the economic return that the owner demands. Even if a client opts for an eighty-year rotation and decides not to have a tree older than eighty years in there, it is still, in my judgment, best to have a multiaged system where you can always have a forest, and you never have to reestablish your forest. In other words, the Stoddard-Neel Approach could work for such landowners. They would sacrifice some ecological diversity, and I would point that out, but we could still manage with our system in a way that would maintain a good fire regime and natural regeneration.

Over the years, we have cut a whole lot of trees while simultaneously growing millions and millions of board feet of timber. On Greenwood Plantation, for example, our records show that we cut over fifty-six million board feet of sawtimber and pulpwood between 1945 and 1995. During that time we had a timber cruise done every ten years, and they show that the standing volume has increased by forty-eight million board feet over those fifty years. We cut a lot of timber and grew a lot of timber at the same time, and you would never know it by looking at Greenwood today. We have maintained all the components of a healthy longleaf pine–grassland ecosystem. So landowners can realize a substantial economic gain from this system, but they have to accept its limitations.[4]

The economics of the Stoddard-Neel Approach do not work like other forestry methods. One of our biggest jobs is convincing the landowner to see the forest as an endowment, not a liquid asset. You start with the principal, you reinvest the interest, and you draw a good-sized dividend. The more money you have in an endowment, the more interest you are going to be able to draw. That is how we treat the forest. We take a conservative return from the timber base while increasing the quality and volume of timber over a long period of time. And as a by-product, we also manage for age, diversity, and beauty.

Mr. Stoddard was able to fine-tune his timber-marking system to perpetuate not only the forest itself, but the quality of the trees and the species. He got a good return for the landowners, and because they were not demanding a heavy return, he grew timber for the future. Unfortunately, a landowner interested in a big return can undo that work very quickly. We might start with one property that had, say, twenty million board feet on it. Forty years later it might have sixty or eighty million board feet on it, and yet we had cut a lot of timber over those decades and made a pretty good return. In the meantime the price went up from twenty dollars per thousand board feet to four hundred dollars per thousand, so there was a tremendous increase in value. We had, in other words, both provided some income and dramatically increased the value of the investment. Well, we have had a few owners—second, third, or fourth generation—who could not stand to see that value just sitting there on the land. They have had other uses for it, and so they liquidated it. Again, we would not liquidate their timber for them, but they never asked us to because they knew we would not do it. Unfortunately, they had an easy time finding foresters who told them what they wanted to hear—that their forestlands were decadent and overstocked, and that they could make more money by cutting out the old growth and starting all over again. And they made a lot of money doing that, at least in the short term. But they also lost a beautiful, functional forest in the process.

The problem, as I see it, is that we live relatively short lives, and yet there is nothing short term about the Stoddard-Neel Approach. Landowners not only need to be patient and conservation minded; they also need to be able to resist the temptation to cash in on favorable market conditions when they come along. If you want to manage for a multiaged longleaf forest, you have to think on a time scale of at least four hundred years. You cannot get around this fact: it takes four hundred years to grow a four-hundred-year-old tree. If an owner wants to bequeath to the future a truly multiaged class forest, he or she will have to leave some trees to grow well beyond their own lives or even the lives of their grandchildren. And that is a hard thing to do, particularly as the value of the land and timber increases over time. Mature standing timber is a mighty temptation for some.

So we are dealing with a long rotation. Greenwood has some trees on it that probably would age out at four to five hundred years old—not many, but there are some of them there. A lot of them are two to three hundred years old. As you get into older age classes, you get fewer of those trees. That is a true multiaged stand, one of the few good ones around. Over at Ichauway, to the contrary, they have some beautiful forested land, but the timber is all second growth, and they do not have many trees that are more than a hundred years old. They have done great work over there, and yet, the daunting fact is that they will need many more years of patient work to get a forest like Greenwood's.

If you want a truly multiaged stand, you cannot clear-cut. You cannot take a thousand acres and develop a multiaged system the first few years. That is one of the primary reasons we try to impress on people not to overcut their timber, particularly if they want to have a multiaged forest. If they want to have a forest and get some income, then they have got to go through the time period of establishing it, which no single individual can do. It takes three generations or more to do that, to get to a perfect longleaf rotation.

We manage for all types of quality—ecological, aesthetic, and economic—and part of how we have been able to do that is by taking a particular approach to the marketplace. Markets can be fickle, and this is a constant challenge, but we try to manage for high-value timber. We sell our timber primarily for quality saw timber, but we get the highest price—besides the specialty market for old-growth longleaf—for poles. The pole market might disappear soon, as other materials begin to be used for poles. The price range runs from poles to saw timber on the higher end, and then there is the combination of sawtimber and pulpwood on the lower end. The difference between a pole and sawtimber is based on specific measurements. A pole has to be line-straight and within a certain diameter range at the top and at the ground line, and it cannot have any type of defect. It has to be a clean, straight, solid tree. For sawtimber trees, you can have ring knots, cronartium blisters, and other defects. It is fairly easy to cut those defects out of a tree and still have good logs. We generally grow for higher-quality markets, and we mark for higher quality, too, by removing most of the poor-quality trees early. When we mark timber, we have to keep our markets in mind.

When we run a pole cut through a tract of timber, it is usually because the stand is too dense and needs opening up. We prefer to take out the lowest-quality trees first, but sometimes you have to cut some good trees, too. The pole market demands some of the good ones. There are times when a gap, along with the sunlight it provides, is needed, and that is what we look for when we mark for poles. But unlike many foresters, we do not cut every potential pole in the forest. We do not harvest a tree just because it reaches pole size. That would not take into account the health of the entire

FIGURE 10. Leon Neel working a pole cut in the early 1950s. Poles were the most desirable use of timber in the Stoddard-Neel Approach because they secured the most return and were suitable for single-tree selection.

system. There is a window of time when a tree is at pole size, but we do not mind letting a few pole-sized trees grow out of pole class. In fact, if we took every pole tree at the right moment for the pole market, we would be selecting for a certain kind of forest, and we are not comfortable with that. So we keep some pole trees and let them age. We can still sell such trees as saw timber later on. This is one of the most important balancing acts within the Stoddard-Neel Approach: you have to keep one eye on the market and manage your forest in a responsive way, but you cannot let the market totally dictate which trees you take and which you leave behind. You need to balance market demand, which shifts all the time, with your ecological goals.

Of course, all of the marking, cutting, and selling of timber I have discussed up to this point happens in a fairly mature forest, usually dominated by longleaf. But it is also important to keep in mind that many of our forests have mature loblolly and slash pine as well. We do not work exclusively on longleaf lands, though we do have a preference for them. We use the same principles on other pieces of land, but loblolly and slash demand some different fine-tuning. The biggest difference is in our use of fire. Loblolly and slash do not have the resistance to fire that longleaf does, especially when the trees are young. Once they are beyond twelve to fifteen years old, we burn loblolly and slash stands just as we would longleaf, but then that makes it difficult to encourage reproduction. That is what makes a stand of loblolly or slash so challenging: it is difficult to have a multiaged forest that is sustainable over a long period of time. Even if you exclude fire to capture reproduction, you will choke out the ground cover and have a dense forest that is not only poor wildlife habitat but also a fire hazard. Ideally, we would like to gradually convert those stands to longleaf over a long period of time, largely because longleaf is so much better adapted to the site management practices of Stoddard-Neel.

Our entire system of marking and cutting timber hinges on the quality of the logger doing the work. The logger is just as important as any other factor in the Stoddard-Neel Approach, and it is getting more difficult these days to find a good, conscientious logger. A lot of landowners allow the logger to do the selection, which is a bad idea. There are a handful of loggers that I might trust in a stand of planted pines, but I have never just turned them loose and told them they can take a certain number of board feet. Also, you have to be very careful to make sure logging does not damage your ground cover too substantially. If it does, hardwood dominance can creep in there quite quickly. So we do our best to minimize damage to the grasses and other ground cover, and to concentrate logging activities in areas that have already been subject to some disturbance.

The main problem we have now, ecologically speaking, is the size of the equipment and the carelessness with which it is used in the woods. Skinning trees, running over reproduction, tearing up ground cover, even felling trees—you can throw a tree one way and do no harm but throw it another way and do a tremendous amount of damage. I learned from Mr. Stoddard that you have to keep close watch over the loggers, and that takes time and energy. He had total control of the logging crews that worked the lands he managed, and his contracts with sawmills detailed exactly what was expected of the loggers. I have probably emphasized good, clean logging even more than he did. I never have shied away from telling loggers how to conduct their business on a piece of land under my managerial responsibility, because that was understood in advance. We have used the same crews over the years, and it has been really hard to bring in new ones.

There has been a tremendous change in the logging business since I began working in these woods. When we first started, loggers worked with crosscut saws and mules. One or two of them had a small farm tractor that they had just started using, but mostly they were working by hand. And it showed. The more you can do by hand, the better the logging job is, because you can move through the woods without destroying as much. But it has gotten now so that a man cannot be a logger unless he has a million and a half dollars worth of equipment, and that is on the low end. These machines have tremendous tires or tracks, and it takes a lot of ground to turn them around. This, in turn, affects how people mark their trees, because they have to open up the forest enough to allow the machines to maneuver. We have not succumbed to that yet, but a lot of people will cut more trees just to make room for the equipment. If they do not cut them, the machines are going to damage the trees, and then you have to cut them anyway. And besides the damage those big machines cause, the operators need volume to make their machinery payments every month, and then try to make enough over that to live. So there is serious economic pressure to overcut these forests coming not only from the landowner, but from the heavily capitalized loggers as well.

We have had some extraordinarily good loggers to work with us over the years, people who took an interest in what they were doing and an interest in what we were trying to do. They learned how to work within our guidelines and run their businesses so they could make enough to turn a profit. We never begrudged them that. But when loggers, like some landowners and foresters, seek to maximize their short-term profits, the forest suffers—and ultimately they suffer in the long term also, because there are fewer trees to cut in the future. Creating the mechanisms and incentives to care for these forests on a long-term basis is one of the most formidable challenges we face in trying to protect and responsibly manage them. And to do it right, there has to be an incentive for everyone who has anything to do with the forest, from logger to forester to landowner, to care for the future of the forest. A serious problem in any sustained yield method of forest management is that the life span of a tree, especially a longleaf pine, is several times that of a forester or logger or landowner, so it is difficult for some to leave valuable older trees in the stand for the future.

It should be clear by now that maintaining the Stoddard-Neel Approach has been an uphill battle over the years, but we were able to do it because we had receptive landowners. They wanted a good aesthetic and recreational experience in a beautiful forest, and a modest return to help them keep it up. But that has gradually changed. One of the major problems I have faced here in the Red Hills over the last couple of decades is the relatively new landowner goal of quail maximization at the expense of everything else. I know I can manage for quail and maintain a healthy forest: we did it for years. But these days, it is tough for me to talk about quail management, because so many people focus on quail to the detriment of the rest of the system. They treat quail more like a crop, the same way foresters have treated timber like a crop. Rather than creating a quality quail habitat in a distinctive and diverse woodland environment, they are out for quantity, and that means simplifying the environment or trading off the ecosystem quality they would get under the Stoddard-Neel Approach. They are deemphasizing the entire experience of the hunt so that they can boast about covey numbers.

A lot of it has to do with ego. There is tremendous pressure among preserve owners today to brag about the number of quail they have. So quail people tell them if they want to maximize their quail population, they can only have so much timber standing on the land. Again, quail are early successional birds, and the longleaf-grassland forest is an old, fire climax system, so they do not think a landowner can have both. But wise management, which includes fire, can balance the early successional habitat of quail while maintaining the forest as well. You do not have to plow up native ground cover or cut most of your timber to get quail habitat.

Since we have been in business, the quail plantations have gone in several directions. Some have taken our system and turned toward real ecosystem management; some continue to focus on quail, but keep a concern for the health of the forest; and some have turned exclusively to quail management and have cut most of their timber. The latter rely on quail research as an excuse to reduce their timber, but what many really want to do is justify selling all their timber. It is private land, and they can do whatever they want to with it within reason, but liquidating timber is not conducive to good conservation in my judgment. It is shortsighted management.

If you bought a five-thousand-acre piece of land, and you wanted the highest number of quail and expense was no problem, you could simply clear the property of all its woodland. You could put in a series of small fields—and I do not mean one here and one yonder. I mean occupy the entire property with a series of fields, field edges, and fallow strips. You could have 50 percent of the land in the cultivation of small fields and accompanying field edges, hedgerows, and escape cover. You would have nesting and escape cover all intermingled, emulating the midwestern farm landscape of the mid- to late 1800s. You might have a landscape like that favored by Adam Bogardus, the famous sportsman who wrote the book *Field, Cover, and Trap Shooting* in 1874.[5] He spent most of his time in the Midwest, but he once came into Alabama and Georgia and shot some quail at somebody's invitation. He went back and told folks he would not go down there again; he could kill more birds in Michigan and Wisconsin and Ohio shooting the hedgerows. He was talking about killing several hundred quail a day. Well, if that is what you want, you can create a landscape that might produce that many quail. But from an ecological standpoint, that is not helping anything.

The difference between such efforts at quail maximization and the Stoddard-Neel Approach is a difference in attitude; it is a difference in land stewardship philosophy; it is a difference in the obligation that some of us feel to not destroy the world in our lifetimes. I am happy to manage for quail, but I also try to get landowners to see that they can have so much more than quail. They can have a sustainable timber crop; they can have beautiful forests in which to hunt; they can have an inexpensive management tool in prescribed fire; they can have a diverse understory that supports the full evolutionary expression of the longleaf-grassland system. I think it is worth it, both for them and the ecological health of the region. But the trick is convincing landowners and hunters. Part of what the Stoddard-Neel Approach aims to do, then, is to make the hunt itself an ecological learning experience. Like Mr. Stoddard's good friend Aldo Leopold once argued, we need to get away from the trophy mentality that has affected so much hunting and realize that, ultimately, good hunting is about a deeper recreational experience where ecology and aesthetics converge.

I am deeply concerned about the future of our longleaf environment. We have tried to legislate conservation measures, but I do not have much faith in political solutions. I finally decided several years ago that I did not have much influence beyond the land I had control over. As long as I had managerial control over it, and the owners were satisfied, I had some good land, but once I lost an owner, the land usually lost its long-term productivity and integrity, and there was nothing I could do about it. Private landowners can still largely do what they want with private lands.

For a time, I put a lot of hope in conservation easements, but they are being ignored by a lot of people at this point. Conservation easements involve landowners selling the rights to develop their land to a conservation organization, yet still retaining ownership of the land itself. Such an arrangement seemed to promise the best of both worlds: private ownership with limitations on management and development that provide landowners with some tax relief. I have written several conservation easement management plans for plantation lands in the Red Hills, which I hoped would provide ample protection for them. But I have seen these easements ignored time after time.

There is a lack of enforcement, but the problems with easements run a great deal deeper than that. With a conservation easement, you write out a clear plan for how you are going to manage the land, laying out your long-term goals, and you legally restrict what you can and cannot do on that piece of land. The problem is, if the easement ends up in the wrong hands, it can be completely corrupted. I have one easement not far from my home that I worked on for years. I wrote the forest management plan for the land, and it was a good, solid, conservative easement. It outlined a series of what we called "special natural areas." Those could be any number of things, such as a little one-acre bog with a wide diversity of plants—the diversity that makes the world so unique and great. As far as the timber was concerned, I had built a tremendous timber volume on that land over the years. We had red-cockaded woodpeckers in there in natural cavities, and we had been increasing the woodpecker population since 1950. We drew this easement up so that the owner could only cut a specific portion of the forest growth increment. I wrote an easement that I felt would be easily understood and would allow successive owners to cut some timber while also protecting the timber base and the ecological values that flowed from it. Well, the landowner and his wife died, the children inherited the land, and they decided to sell it. Before the new owner bought it, their lawyers checked everything out, including the easement. They knew it had an easement on it, and they knew what the easement said. To cut to the chase, the new owners have completely ignored the easement over the last several years. They have created a tremendous amount of land disturbance over there, including cutting a volume of timber that far exceeds what they are allowed to do under the easement. And the organization that holds the easement has done nothing to stop it. So the easement is ultimately no good and the damage is done. Experiences such as that one have shaken my faith in easements.

At one time I used to say—and this was a long time ago—that the only way you could perpetuate the environment was through private ownership. I did not think much of public ownership, because we had such sorry public management years ago. And we tended to have conscientious private owners. But I have changed my mind on that. If there is any hope now, it is probably through public ownership of at least some of these critical longleaf lands.

Some people say the Stoddard-Neel Approach cannot work for the majority of landowners in the longleaf belt, but I believe it can. There are landowners out there who have a conservation ethic and do not want to liquidate their timber, and I am hopeful that their ranks will grow. Our system can help them. Our philosophy works the same on a small tract as it does on a large one. We want to leave some for the future. That is the whole point of the Stoddard-Neel system: our historic longleaf forests took hundreds of generations to develop, and the values that inhere in the best remnants of those forests will last more than one generation. Indeed, effective longleaf restoration, which more and more people have begun to embrace, will also be a multigeneration process. Our forestry must look beyond one generation as well.

AFTERWORD

The Legacy of Leon Neel

by Jerry F. Franklin

The incredibly rich and complex longleaf pine ecosystem of the southern coastal plain is without parallel in the diversity of its ground cover, with hundreds of species of herbs and grasses often present within a single stand, sustaining an incredible array of wildlife—birds, mammals, reptiles, and amphibians—with some, such as the red-cockaded woodpecker, functioning as ecological flagships. Perhaps mostly importantly, the longleaf pine–grassland system is a temperate forest type unique in the degree to which it is attuned to, and dependent on, a hyperfrequent cycle of fire. It is one of North America's extraordinary forest ecosystems. And for over sixty years, the longleaf ecosystem has had in Leon Neel a forester who has understood its richness and complexity and

FIGURE 11. Leon Neel and Jerry Franklin became acquainted while they served on the Scientific Advisory Committee of the Joseph W. Jones Ecological Research Center at Ichauway. Franklin is an architect of the "new forestry," an ecosystem approach to forestry that shares many philosophical principles with the Stoddard-Neel Approach.

developed and demonstrated a management approach that maintains its integrity.

Leon Neel is a forester who was ahead of his time but always in tune with his roots—the longleaf pine ecosystem. His is an extraordinary story of stewardship, knowledge born of experience, and the courage of convictions in the face of a profession that long had no place for the stewardship of an ecosystem. His love and his understanding of the longleaf forest required him to go against the flow of his profession, which was focused on the goals of maximizing economic wood production even at the cost of essentially all of the natural elements of the pinelands.

Those associated with the corporate wood products industry, committed to ever-higher levels of economic efficiency, have never appreciated Leon Neel's management approach. Nor have the academic centers, which have been bent on producing technicians and knowledge in the service of corporate goals. Quantity, not quality, was what these groups desired. And the best way to produce wood pulp in quantity was to replace the longleaf pine woodlands with an entirely different kind of ecosystem—even-aged plantations of faster growing pine species such as slash or loblolly. In the process of achieving this violent transition, one of the most profound and important in the recent environmental history of North America, the conservation values and biota of the longleaf pine–grassland system were marginalized or eliminated—they had to either adapt to high-yield forestry or be lost.

Yet, during this same era, a different, contrasting approach to forest stewardship emerged in the longleaf pine region—the Stoddard-Neel Approach. In fact, the Stoddard-Neel Approach was the pioneering effort in the development of ecological forestry—forestry based on ecological principles and a thorough understanding of the ecosystem in question. It is also an approach to forest stewardship that allows for the integration of environmental, economic, and cultural objectives. As such, it is worthy of our close attention, because of both its historical import and current applicability. Indeed, it is amazing how relevant the Stoddard-Neel Approach and the concepts that it incorporates have become as we enter the twenty-first century. Not only was the Stoddard-Neel Approach ahead of its time then, but it is remarkably well positioned to help us meet the challenges that we face.

Where are we today in forestry? The corporate forestlands are fading fast, and the future of the southeastern pulp industry as a whole is uncertain. Globalization of the world's markets has allowed the fiber farms of the southern hemisphere to eclipse the North American wood products industry as a source of common wood fiber. They have taken the concepts of high-yield forestry—pioneered in the southeastern United States—to ever more intense, technical, and efficient levels, marginalizing the United States in the global wood market. As a result, the pinelands of the Southeast face a critical moment of transition.

Foresters and land managers now find themselves living in a society that wants its forests to be managed for an array of values, not just wood. Americans increasingly care about the rich diversity of plants and fauna, rare orchids, and the red-cockaded woodpeckers, as well as the quail and the deer found in the longleaf pine ecosystem. Society also cares about the role that forests play in maintaining a well-regulated flow of high-quality water; the vast majority of our municipal water supplies come from forested watersheds, and we are coming to recognize and value the broader range of ecosystem services that they provide. Other concerns are the important role of forests in storing carbon and the potential role of forests as a renewable source of energy. And, of course, our society values forests for their recreational and inspirational values. With such wholesale changes in the southeastern forest products industry, then, we have a remarkable opportunity to re-envision what we want from the region's woodlands, and the rich experience of Herbert Stoddard and Leon Neel shows us how we might best manage them to achieve such results. Future forestland stewardship must be about integrating a broad array of forest values rather than marginalizing most in favor of a single dominant use. This is what Stoddard-Neel is all about!

Ours is also an era of great uncertainty with regard to our forests. Climate and land-use change threaten the integrity and, in some cases, even the continued existence of forested lands. Such fundamental changes in both the environment and society are occurring rapidly, and it is difficult to predict the consequences. How is climate change affecting forest health and function? Where will markets move with regard to forestlands and forest values? One approach to dealing with such uncertainty would be to manage forests in ways that will reduce risk, increase future societal options, and increase ecosystem resiliency. These are very different emphases than those that are found in traditional production forestry. But, once again, they are what the Stoddard-Neel philosophy of management is all about.

How fortunate we are to have this philosophical and practical model of ecological forest stewardship at this critical juncture. Thanks to Leon, it has persisted through the flood tide of plantation forestry to provide us today with an example of how forests can be managed so as to sustain the complexity of the ecosystem, economic returns, and cultural values. Leon Neel and his mentor, Herbert Stoddard, understood intuitively that attempting to maximize any single output—whether it be wood or quail—would eliminate or marginalize other forest values. The focus of their stewardship has long been on ecological integrity and a deeply rooted conservation ethic—on "keeping all of the parts," to paraphrase Aldo Leopold, who was himself influenced by Herbert Stoddard's pioneering work. They have insisted that what was left behind from disturbances, both natural and managed, is what enriches future stands.

It seems as though in each major forest type a forester emerges who is known for the breadth of his knowledge of the natural history and practical management of that type—a "forest guru" so to speak. Certainly, Leon Neel has attained that status for the longleaf ecosystem, based on a lifetime spent observing and caring for these forests. By doing so he has provided us with both a powerful management philosophy and a demonstrated approach to the application of that philosophy. He has also embodied for us the virtues of personal integrity and dedication to the core principles of resource stewardship. Thank you, Leon Neel!

NOTES

INTRODUCTION
Forestry beyond One Generation

1. On Stoddard's background, see Herbert L. Stoddard, *Memoirs of a Naturalist* (Norman: University of Oklahoma Press, 1969); and Albert G. Way, "Burned to Be Wild: Science, Society, and Ecological Conservation in the Southern Longleaf Pine" (PhD diss., University of Georgia, 2008).

2. On sharecropping in the South at large, see Pete Daniel, *Breaking the Land: The Transformation of Cotton, Tobacco, and Rice Cultures since 1880* (Urbana: University of Illinois Press, 1985); and Jack Temple Kirby, *Rural Worlds Lost: The American South, 1920–1960* (Baton Rouge: Louisiana State University Press, 1987). On the Red Hills specifically, see William Warren Rogers, *Transition to the Twentieth Century: Thomas County, Georgia, 1900–1920* (Tallahassee: Sentry Press, 2002); Clifton Paisley, *From Cotton to Quail: An Agricultural Chronicle of Leon County, Florida, 1860–1967* (Gainesville: University of Florida Press, 1968); and Way, "Burned to Be Wild."

3. For a description of the hunt, see George M. Humphrey and Shepard Krech, *The Georgia-Florida Field Trial Club, 1916–1948* (New York: Scribner Press, 1948).

4. Herbert L. Stoddard, *The Bobwhite Quail: Its Habits, Preservation, and Increase* (New York: Charles Scribner's Sons, 1931); Way, "Burned to Be Wild." For more on the history of wildlife management, see Thomas R. Dunlap, *Saving America's Wildlife: Ecology and the American Mind, 1850–1990* (Princeton: Princeton University Press, 1988); Curt Meine, *Aldo Leopold: His Life and Work* (Madison: University of Wisconsin Press, 1988); and Julianne Lutz Newton, *Aldo Leopold's Odyssey* (Washington, D.C.: Island Press, 2006).

5. *The Fire Forest: Longleaf Pine–Wiregrass Ecosystem* (Covington, Ga.: Georgia Wildlife Press, Georgia Wildlife Federation, 2001). For a good treatment of the history and ecology of the longleaf forest, see Lawrence Earley, *Looking for Longleaf: The Fall and Rise of an American Forest* (Chapel Hill: University of North Carolina Press, 2004). On one of the

earliest fire advocates, see Elizabeth Finley Shores, *On Harper's Trail: Roland McMillan Harper, Pioneering Botanist of the Southern Coastal Plain* (Athens: University of Georgia Press, 2008).

6. Timothy Silver, *A New Face on the Countryside: Indians, Colonists, and Slaves in South Atlantic Forests, 1500–1800* (Cambridge: Cambridge University Press, 1990), 12–15; Earley, *Looking for Longleaf*, 18.

7. Roland Harper, "Geography and Vegetation of Northern Florida," in *Florida State Geological Survey, Sixth Annual Report*, ed. E. H. Sellards (Tallahassee: State Geological Survey, 1914); Steve Gatewood et al., *A Comprehensive Study of a Portion of the Red Hills Region of Georgia* (Thomasville, Ga.: Thomas College Press, 1994).

8. E. V. Komarek, "The Natural History of Lightning Fire," *Proceedings Tall Timbers Fire Ecology Conference* 3 (1964): 139–83; Stephen J. Pyne, *Fire in America: A Cultural History of Wildlife and Rural Fire* (Princeton: Princeton University Press, 1982).

9. For the contours of this debate, see Paul A. Delcourt, Hazel R. Delcourt, Dan F. Morse, and Phyllis A. Morse, "History, Evolution, and Organization of Vegetation and Human Culture," in *Biodiversity of the Southeastern United States: Lowland Terrestrial Communities*, ed. William H. Martin, Stephen G. Boyce, and Arthur C. Echternacht (New York: Wiley, 1993), 47–80; W. A. Watts, and B. C. S. Hansen, "Pre-Holocene and Holocene Pollen Records of Vegetation History from the Florida Peninsula and Their Climatic Implications," *Palaeogeography, Palaeoclimatology, Palaeoecology* 109 (1994): 163–76; Cecil Frost, "History and Future of the Longleaf Pine Ecosystem," and D. Bruce Means, "Vertebrate Faunal Diversity of Longleaf Pine Ecosystems," in *The Longleaf Pine Ecosystem: Ecology, Silviculture, and Restoration*, ed. Shibu Jose, Eric J. Jokela, and Deborah L. Miller (New York: Springer, 2006), 9–42, 157–216.

10. W. G. Wahlenberg, *Longleaf Pine: Its Use, Ecology, Regeneration, Protection, Growth, and Management* (Washington, D.C.: U.S. Forest Service, U.S. Department of Agriculture, 1946); Shibu Jose, Eric J. Jokela, and Deborah L. Miller, eds., *The Longleaf Pine Ecosystem: Ecology, Silviculture, and Restoration* (New York: Springer, 2006).

11. Earley, *Looking for Longleaf*, 38–40. At small scales, the longleaf understory rivals rainforest areas in diversity of plant species, though as the scale gets larger, rainforests quickly surpass longleaf forests in numbers of species.

12. Bruce Means, "Longleaf Pine Forest, Going, Going . . . ," in *Eastern Old-Growth Forests: Prospects for Rediscovery and Recovery*, ed. Mary Byrd Davis (Washington, D.C.: Island Press, 1996), 210–29.

13. Herbert L. Stoddard, Henry L. Beadel, and E. V. Komarek, *The Cooperative Quail Study Association, July 1, 1934–April 15, 1943*, Misc. Pub. no. 1 (Tallahassee: Tall Timbers Plantation, 1961).

14. For a history of the first twenty years of Tall Timbers, see E. V. Komarek, *A Quest for Ecological Understanding: The Secretary's Review*, Misc. Pub. no. 5 (Tallahassee: Tall Timbers Research Station, 1977).

15. Many of these debates took place within the U.S. Forest Service. See Paul W. Hirt, *A Conspiracy of Optimism: Management of the National Forests since World War II* (Lincoln: University of Nebraska Press, 1994), 22–25; Nancy Langston, *Forest Dreams, Forest Nightmares: The Paradox of Old Growth in the Inland West* (Seattle: University of Washington Press, 1995); Way, "Burned to be Wild," chap. 6.

16. "Memo of Understanding Governing the Sale and Cutting of Poles on Sinkola Land Company and other Plantations of the Thomasville Section," Stoddard Papers, Contracts, Tall Timbers Research Station, Tallahassee.

17. William Boyd, "The Forest Is the Future? Industrial Forestry and the Southern Pulp and Paper Complex," in *The Second Wave: Southern Industrialization from the 1940s to the 1970s*, ed. Philip Scranton (Athens: University of Georgia Press, 2001), 168–218; William Clarence Boyd, "New South, New Nature: Regional Industrialization and Environmental Change in the Post–New Deal American South" (PhD diss., University of California, Berkeley, 2002); Jack Temple Kirby, *Poquosin: A Study of Rural Landscape and Society* (Chapel Hill: University of North Carolina Press, 1995), 197–234; Way, "Burned to be Wild," Chap. 6.

18. Janisse Ray, *Ecology of a Cracker Childhood* (Minneapolis: Milkweed Editions, 1999), and *Wild Card Quilt: The Ecology of Home* (Minneapolis: Milkweed Editions, 2004).

19. See Earley, *Looking for Longleaf*, for details on many of these efforts.

20. Paul S. Sutter and Albert G. Way, "The Stoddard Neel Method: An Oral History" (2005).

21. Albert G. Way, "The Stoddard-Neel Method: Forestry beyond One Generation," *Forest History Today* (Spring/Fall 2006): 16–23.

22. See R. K. McIntyre, S. B. Jack, R. J. Mitchell, J. K. Hiers, W. L. Neel, *Multiple Value Management: The Stoddard-Neel Approach to Ecological Forestry in Longleaf Pine Grasslands* (Newton, Ga.: Joseph W. Jones Ecological Research Center, 2008); *The Fire Forest: Longleaf Pine–Wiregrass Ecosystem* (Covington, Ga.: Georgia Wildlife Press, Georgia Wildlife Federation, 2001); and R. J. Mitchell, J. K. Hiers, J. J. O'Brian, S. B. Jack, and R. T. Engstrom, "Silviculture That Sustains: The Nexus between Silviculture, Frequent Prescribed Fire, and Conservation Biodiversity in Longleaf Pine Forests of the Southeastern United States," *Canadian Journal of Forest Resources* 36 (2006): 2724–36.

CHAPTER 2
Time Well Spent with Mr. Stoddard

1. James V. Carmichael won the popular vote, but under the county-unit system, Eugene Talmadge won the county-unit votes, thereby making him governor-elect. By the time of the November general election, however, Ole Gene was in ill health, and a group of Talmadge family supporters organized a write-in campaign for his son Herman. Gene died a few weeks before the inauguration, and the Georgia General Assembly had the responsibility to decide who would take his place. Through a duplicitous series of events, including the "discovery" of fifty-six additional write-in votes from Talmadge's home county of Telfair, the General Assembly voted Herman Talmadge governor. Acting Governor Ellis Arnall, however, refused to give up his office and insisted the governorship should pass to his lieutenant governor, Melvin E. Thompson. Talmadge and his supporters physically seized the office and residence, and Thompson set up an office as governor in exile in downtown Atlanta. In March 1947 the state supreme court declared Thompson to be acting governor until a special election in 1948, which Herman Talmadge won comfortably.

2. E. V. Komarek, "Herbert L. Stoddard, Sr.—'The Wizard of Sherwood,'" *The Oriole* 37 (December 1972): 24.

3. Herbert L. Stoddard, *Memoirs of a Naturalist* (Norman: University of Oklahoma Press, 1969).

4. Herbert L. Stoddard, *The Bobwhite Quail: Its Habitat, Preservation, and Increase* (New York: Scribner, 1931).

5. Roger Tory Peterson and James Fisher, *Wild America: The Record of a 30,000-Mile Journey around the Continent by a Distinguished Naturalist and His British Colleague* (Boston: Houghton Mifflin, 1955).

CHAPTER 3
The Early Years of Tall Timbers Research Station

1. Stoddard Field Diaries, May 5, May 7, 1924, HLS Papers, Archives of Tall Timbers Research Station, Tallahassee.

2. Robert L. Crawford, *The Great Effort: Herbert L. Stoddard and the WCTV Tower Study*, Misc. Pub. no. 14 (Tallahassee: Tall Timbers Research Station, 2004), 40.

3. Harold H. Biswell, *Ponderosa Fire Management: A Task Force Evaluation of Controlled Burning in Ponderosa pine Forests of Central Arizona*, Misc. Pub. no. 2 (Tallahassee: Tall Timbers Research Station, 1973); Biswell, *Prescribed Burning in California Wildlands Vegetation Management* (Berkeley: University of California Press, 1989).

4. Stephen J. Pyne, *Fire in America: A Cultural History of Wildland and Rural Fire* (Princeton: Princeton University Press, 1982), 490.

5. E. V. Komarek Sr., *A Quest for Ecological Understanding: The Secretary's Review*, Misc. Pub. no. 5 (Tallahassee: Tall Timbers Research Station, 1977), vii.

6. Ibid., vii–viii.

7. Ibid., viii.

CHAPTER 4
The Stoddard-Neel Approach: Managing the Trees for the Forest

1. See, e.g., R. J. Mitchell, J. K. Hiers, J. J. O'Brien, S. B. Jack, and R. T. Engstrom, "Silviculture That Sustains: The Nexus between Silviculture, Frequent Prescribed Fire, and Conservation of Biodiversity in Longleaf Pine Forests of the Southeastern United States," *Canadian Journal of Forest Research* 36, no. 11 (2006): 2724–36; Steven B. Jack, W. Leon Neel, Robert J. Mitchell, "The Stoddard-Neel Approach: A Conservation Oriented Approach," in *Longleaf Pine Ecosystem: Ecology, Silviculture,*

and Restoration, ed. Shibu Jose, Eric J. Jokela, and Deborah L. Miller (New York: Springer, 2006), 242–45.

2. Herbert Stoddard, "Memoirs of a Naturalist" (original manuscript in Leon Neel's possession), 460.

3. P. A. Seamon, R. L. Myers, L. E. Robbins, and G. S. Seamon, "Wiregrass Reproduction and Community Restoration," *Natural Areas Journal* 9 (1989): 264–65.

4. For Greenwood statistics, see R. K. McIntyre, S. B. Jack, R. J. Mitchell, J. K. Hiers, and W. L. Neel, *Multiple Value Management: The Stoddard Neel Approach to Ecological Forestry in Longleaf Pine Grasslands* (Newton, Ga.: Joseph W. Jones Center for Ecological Research at Ichauway, 2008), 24–25.

5. Adam H. Bogardus, *Field, Cover, and Trap Shooting. Embracing hints for skilled marksmen; instructions for young sportsmen; haunts and habits of game birds; flight and resorts of water fowl; breeding and breaking of dogs* (New York: J. B. Ford, 1874).

INDEX